D0169724

PONIES + HORSES BOOKS
Published by Ponies + Horses Books
Ponies + Horses Books (Scotland), 7b The Hidden Lane, 1103 Argyle Street, Glasgow, G3 8ND
Ponies + Horses Books (Canada), Box 271, Stn C, Toronto, ON, M6J 3P4
www.poniesandhorsesbooks.com

First published by Ponies and Horses Books in 2016

Photography Credit: Darla Flowers of Photos by Darla
Photography Credit: Carole Tothe-Gurgol of Tothe Photography
ISBN: 9781910631706

Beamsville, ON

CHEERS TO
BIG DREAMS AND
DELICIOUS WINES!

CHRISTINA
BROWN.

to build a vineyard

an adventure of love, wine + courage

Christina **Brooks**

BACK 10 CELLARS

EST. 2002

For my husband, Andrew, who has been by my side on this most wonderful journey and continues to love me unconditionally... and for Ellery and Amelia, my life's most precious cargo and the greatest loves of my life.

It's all for you.

"Too many of us are not living our dreams
because we are living our fears."

Les Brown

Contents

1999

A visit to Burgundy

2002

Renovating Main House

2005

Clearing fallow orchards

2008

Building the office

2013

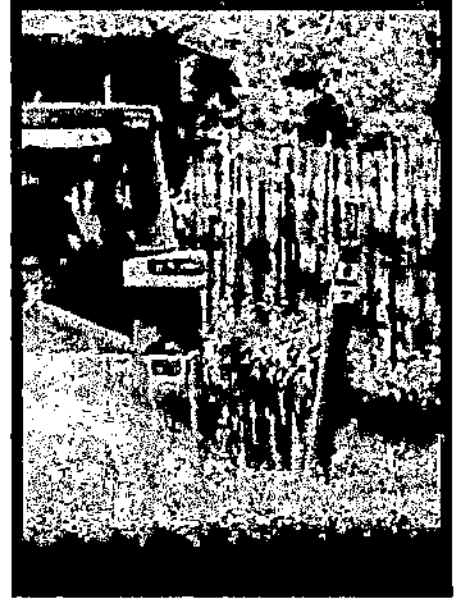

6 year old vines

2016

and beyond...

1. The Big Leap

My heart pounds and sweat drips into my eyes. It feels like 100 degrees out here and the humidity has turned me into the worst version of myself.

"Andrew! Andrew!" I shout at the top of my lungs, hoping he can hear me through the open bathroom window.

"Help! Back! Back!" I threaten this goat with matted, muddy fur and blood red eyes.

He is hell-bent on eating my flat of fresh flowers that's waiting to be planted and I'm not having it. My rake pokes at the air, keeping him at a safe distance. He snorts, looking at me sideways, dragging his hooves back and forth on the dusty gravel driveway, forcing a dry smoky dust into my eyes. I move left. He moves right. A dance between woman and beast.

Andrew runs out with one hand holding his towel and the other cupped around his lips like a trumpet.

He screams, "Get out! Get out!"

His hair is sopping wet and still soapy in spots making some of his brown waves cling to his forehead above his furrowed brows. He is mortified by the sight of his wife trying to hold off a wild beast in the driveway of our new vineyard.

"Get! Get!" he yells louder, using one hand to wave threateningly at my aggressor, the other gripping his slipping towel.

Red Eyes retreats reluctantly and we burst into laughter. Andrew sits on the stair of the back stoop, half naked and shakes his head from side to side, an enormous smile unfolding. My giggles increase and I laugh so hard now that I cry, tears streaming down my face. Despite the brief moment of giddiness, I am emotionally drained and exhausted from the first few months of our new adventure.

An aggressive goat is the final straw and I just want to curl up with an extra foam latte and pretend none of this is happening.

"We just bought a rundown vineyard in Niagara," we told our friends in Calgary when the deal was finalized that January. Their reaction was always the same. The jaw drops, the heads slowly shaking from left to right and then the disbelieving, stuttering exclamations, "Really? Wow, that's, errr... great!"

"Yep, we did it!" Andrew would say confidently, with me being the only person who could detect that slight hint of fear in his voice.

"There's no turning back now," he would add with a smirk and a nod.

Looking back now, I realize that their looks of horror came from a loving place of concern. We were only 32 and 28 and when people that age cash in their lives and quit their jobs to chase a dream, sure, it sounds romantic, but it can also be utterly devastating if things go awry. Every time we told the story I felt as if I was not only trying to convince my friends of this imminent ascent into blissful vineyard ownership, I was trying to convince myself. The words slipped effortlessly from my mouth, but my brain clung to shards of uncertainty and self-doubt.

We had both spent a decade waiting tables and learning about wine in higher end establishments in Calgary. We squirreled away our money and traveled the world and when 911 hit, we began questioning our lives and lifestyle. If we won the lottery, we told ourselves, we would buy a vineyard. So, we hatched a plan. I had my real estate licence by then and we began investing in properties to fix and flip. We were in our early twenties when we bought our first place. We bought rundown houses, made them pretty and rented them out, or flipped them quickly, allowing us to roll cash into the next project. We waited tables by day and renovated houses by night. It was gruelling work. We mapped out our lives on a calendar and told ourselves that if we played our cards right, in a short time, we could sell some of the properties we had been hanging onto and flip another house or two. With the right planning, we could be living the life of our dreams.

It was a little over a year later that we bought our rundown vineyard in Niagara, where I was born and raised. While it was a new side of the country for Andrew, it was a return home for me.

We decided to drive east to Niagara after we had packed up our lives in Calgary. Our SUV was packed to the rafters with what remained of our lives' belongings. Our couches and beds were sold and now only the last piece of furniture remained at the house for the new owners. We would have to start from scratch when we reached Ontario. I laid eyes on the solitary armoire and said one last goodbye, taking a last walk through the abandoned house. Andrew carried out the last of our belongings and thoughtfully opened and closed each kitchen cupboard door looking for missed items.

"Goodbye. You've been good to us," he said, to the very first house we had bought together.

Our house of firsts; where we got engaged, where we brought home our first pets and where we really found ourselves. This place housed the most beautiful journey of our self-discovery.

"We had so many good times here," I said, a little disconsolately.

"I will never forget this place," Andrew said, handing me the keys, choking back tears.

He looked longingly at the front door and ran his hand down the cool copper front one last time with a wistful smile. The floors creaked and the house echoed starkly as I walked through the rooms, drew the drapes one last time and closed the doors on the house I loved, friends I would miss and the life I was leaving behind.

The drive did not turn out to be quite the romantic trip we had envisioned. Our cats, Kenya and Bali, teetered in their travel cages in the back, perched precariously on top of boxes filled with kitchen gadgets.

We had been driving less than two minutes when I felt the 4x4 lurching violently from side to side.

"What are you doing? Slow down!" I yelled to Andrew as I clung on to the door handle of the whipping vehicle.

He looked horrified as he gripped the wheel, grimly, and blurted, "It's not me doing that! I'm losing control of the car!"

A swift stop and a quick phone call to his brother confirmed that we had a "speed wobble." The weight of the U-Haul we were towing was just too much and when its contents lurched back and forth, it threw the car from side to side. We continued at a snail's pace with hazard lights blinking. My nails were digging into the dashboard after two more terrifying wobble incidents and Bali, the smallest of the cats, hadn't stopped groaning an awful visceral half meow, half grunt since trailer first started lurching.

I called the vet from the road.

"The medication we gave them was supposed to relax them, but I think he needs another half pill. He won't calm down and now he's foaming at the mouth."

I looked back at poor Bali and saw a foam beard forming on his little black chin.

The vet said, "Well, this is a first," offering us little comfort.

Bali's eyes were now rolled back and he was sitting up, leaning against the sides of the cage, like he was in his death throes. He looked like a mini furry zombie Santa Claus and the lump in my throat was growing, as my guilt about putting my beloved pets through such torment grew.

We made it as far as Saskatchewan, 720 grueling kilometres from where we started, until I couldn't take it anymore.

In Saskatchewan, we ditched the U-Haul, hired a mover, put the cats on a plane - with a dear friend picking them up from the airport at the other end - and continued on our harrowing journey. One blizzard and four days later, we arrived in Niagara before the sun set, drained but eager to begin our new adventure.

We got the keys to our new property and our new life in March. Andrew carefully slid the rusted-out skeleton key in the lock and the old wooden door creaked open.

It smelled of mould, old leather and damp wood. The floors were covered in pats of dried mud, little stones and large boot prints. The walls were a repulsive grayish purple, like mouldy grapes long forgotten in the back of the fridge, and the faux stone wall to the east of the kitchen looked worse than I had remembered. The stones somehow looked more plastic and they had ghastly, glittery bits that caught the sun, making them impossible to ignore. The windows were cloudy and coated with a layer of grime, making everything outside look blurry, as if seen through fogged-up goggles.

The living and dining room windows were large and thin, and vibrated loudly whenever a car or truck drove by. The fireplace at the end of the living room was made of cracked, old stone with a cement gray colour and a chipped façade. It looked sad, neglected and unloved and I wondered how long the house had been vacant.

"Welcome home," Andrew smirked, and I felt hot, burning liquid crawl up my esophagus.

We creaked up the stairs to the bedrooms and silently glared at the offensively orange shag carpet marked with dark stains. The words "this all needs to get ripped out" were just understood between us without ever having to be spoken aloud.

Most of the doors appeared to be in good shape, only having suffered a scrape here and there, with the exception of the interior of the bathroom door that looked like it had been attacked by a trapped, rabid animal. Large, vertical claw-shaped scrapes had hacked pale scars down the dark wood. I bent and ran my fingers down the rough lines, feeling the prickly, splintered wood and could instantly feel the fear of the animal that had caused them.

We saw ladybugs crawling everywhere in biblical proportions, hopping through shag rugs as if they were obstacle courses, crawling from inside window frames up the walls, down the walls and falling from the popcorn ceiling into the sinks and toilets. The walls were alive with these tiny, crawling black bugs in their black and red polka dotted armour. We stood with our mouths agape watching them. There seemed to be no end to the constant stream of them, parading unapologetically like soldiers marching into the light from their winter hibernation.

It's no wonder that the lady from the bank who came to appraise the house in November took a look around, and then said in a matter of fact tone, "This house needs to be ripped down. The bank won't approve a mortgage on a tear down."

We must have looked like fresh picked daisies smiling naively at her with nothing but hope in our grins. She tiptoed through the house in her pointy heels, sour faced, picking her way around the muddy footprints and dirt like she was avoiding land mines.

"Surely it's not that bad?" Andrew said with some concern. "It just needs some spit and polish."

I chimed in, "It's just a diamond in the rough. She just needs a little work."

The lady from the bank pursed her lips while, eyed the living room and dining room with a dreadful look and didn't respond.

The house was a grand old dame, forgotten for years, and, yes, she needed more than eye shadow and lipstick, but I was optimistic that we could bring her back to her former glory. At least I had thought I was...

The view from the bathroom framed out our newly acquired ten-acre property like a postage stamp. Old fruit trees bending sideways, frayed dangling limbs, sun-bleached grass reaching for the sky, and a desolate orchard and vineyard that looked more like a graveyard than something that should be alive. I shuddered to think of what might be living or dead out there.

The two barns behind the house looked like skeletons, their interiors swept away by time and neglect, barely standing, just hanging on to their foundations. Weeds and mice had taken refuge in their cracked cement floors in their droves. Their doors dangled like loose teeth. Everywhere we looked we saw ominous tasks and time, and money and more tasks. My chest tightened and my breath shallowed.

"What have we done?" I couldn't stop myself from saying out loud as I walked around the property line, regret tugging at the hems of my confidence.

I locked up the house, pretending that the knots in my stomach weren't growing. On the drive to my grandmother's house Andrew and I said nothing to each other. He gripped the steering wheel more tightly than necessary, staring intently straight ahead. I knew he wanted to say something but he just couldn't find the right words. My mouth was dry and pasty like I'd eaten a box of chalk. I couldn't get my mind to slow down and every time I tried to compartmentalize what needed to be done, room by room, my brain betrayed me by throwing the entire, massive project at me, all at once, like a tidal wave of impossibility. I was nauseous and overwrought with panic, desperately wanting to click my heels and go back to my life in the city, in Calgary.

"It will be okay," I told myself, "It will be okay." But I hardly believed it.

My grandmother leapt up from her chair, looking happy to see us. She had a small frame, gentle blue eyes and wiry, grey hair that had replaced the sweeping red beehive she donned in her youth. She walked in quick short steps and seemed to bounce with the energy of a teenager. She was legally blind but her face still lit up when she looked at me. When our eyes met, I always wondered how much of me she could really see.

She was my beacon of light in this madness and when I looked at her on that exhausted afternoon, I saw her love and a sense of calm ensued. She had graciously invited us to stay at her place while we renovated our house and settled into our new lives. I think we must both have looked defeated because no sooner had we walked into the house than she offered us a cold beer with more than a tinge of concern in her expression. I took two big gulps of the icy cold liquid and then excused myself for a rest. I closed the door of the upstairs bedroom, turned off the lights and lay there looking up at the ceiling, into nothingness, trying to steady my breathing. I faded off into a deep sleep and awoke to the smell of something warm and comforting wafting from the kitchen.

My grandmother set down her simple but delicious salad in front of me. I was groggy, but salivating. She knew where everything was in her kitchen, right down to her salt and peppershaker, feeling for the shape of things she needed in the cupboards. This night, she had placed in front of us a small bowl filled with the crisp green iceberg lettuce leaves that I somehow only ever ate at her house and finely diced purple onions. Three fresh tomato quarters were piled neatly like eggs in a bird's nest. I sprinkled a generous amount of Parmesan cheese on top and drowned the contents in Italian dressing, suddenly famished. We feasted on homemade sausages and baked potato wedges ladled in butter and sprinkled with flakes of green followed by her homemade fluffy and delicious apple pie. Cinnamon, a sweet buttery crust and slippery warm apples filled my mouth, relief flooded my body. My grandmother's home became our refuge at the end of each long, hard day. She fed our bodies with food and our spirits with hope.

2. Start from Scratch

We manage to pull the last of the curvy Italianate iron bars from the windows off the front of the house and our new abode begins to look like a house in transition. Where the bars were glued and fastened to the white siding, they leave deep blue round stains that look like bruises on an elderly person with thinning skin. Andrew grabs the bucket filled with hot water and soap and we work on scrubbing them off with careful up and down motions. I step back to take a look at the façade without its bars and the house looks like it is weeping. The windows look like eyes, mascara dripping from its saddened sashes, and a pang of the house's sorrow raises the little hairs on the back of my neck.

"I smell like a dead person," I say to Andrew and he laughs. "Oh God, look at my hair!" Andrew says with mock despair.

He whips off his baseball cap and the shape of his hair

doesn't change. It looks like it's welded on with sweat. The hair on the top of his head is matted, while the curls on the sides near his ears peak upward.

We've been hacking at the waist-high weeds on the front lawn of the property and hauling it all to the back of one of the barns almost fifty feet away, stacking this unruly growth in neat little piles waiting to be burned.

The days develop something of an accidental routine. Work inside for a few hours, break for lunch, work outside for a few hours until it gets dark and, then head inside and work until our bodies can't move anymore or until hunger takes over.

Each day I check fifty things off our mile-long list of things we need to do to make the house liveable. Every task accomplished gets checked off... then new tasks seem to jump to the top of the list out of nowhere. It feels like a never-ending game. I wear a pair of navy workman's overalls and a white mask, my seemingly permanent new wardrobe. I feel the opposite of feminine most days and am desperate to put on something nice and get my hair out of this perpetually knotty ponytail.

Andrew's blue coveralls match mine, but with more muddy handprints, and his work boot soles are splitting apart from the steel toes so that they scoop upward like Dutch wooden clogs.

We pull old rusty nails from the walls that hold up nothing but dirty outlines of pictures that once hung there, those empty squares and oval shapes like hollow reminders of a happier time when people lived in and loved this house. I wonder which pictures used to hang here and which faces smiled back from the walls over time. We fill in the tiny black holes with putty and each downward stroke of the sticky white filler offers the walls a fresh new protective bandage to hide its scars. We tear out smelly, old rugs that fill the rooms with thick speckles of dust when rays of sunshine dance through the windows.

Mornings begin at 7:00 a.m. with the beep of the alarm clock and my near instant awareness of all the knots in my stomach. Friends arrive to the house every few hours with smiles and fast food burgers in tow. They are welcome reprieves, as are the small cartons of bite-sized donuts washed down with copious amounts of coffee that we treat ourselves to, to keep up our dwindling energy.

It's so strange not having an income while renovating this house. I feel displaced, like I'm somewhere that I'm not supposed to be, doing something that I know I'm not supposed to be doing. It feels both a bit self-indulgent and a bit foolhardy at the same time. I keep reminding myself that at the end of this renovation, we will move in, call this house and vineyard our home and begin to focus on making money and a livelihood, but what makes it difficult is the dwindling money in our bank accounts. I log into our bank account almost daily expecting there to be the same amount of savings as the day before, but disappointingly that never happens. We are haemorrhaging money and it's a race to get finished before our savings run dry. Andrew is a silent worrier. When he's worried he gets quiet. When I worry, I just keep talking until I get a response which is usually him looking down at the floor, nodding his head so that I know he can hear me. Andrew cashes in another big hunk of our savings to get us through the next few months.

"I know I will be able to sleep better at night," he confides.

Ironically, it's digging into our savings that we promised not to touch that is now keeping me awake at night.

The back door swings forcefully open and behind it there's an incensed man shouting, "The goats! The goats! They eat all my grapes!"

I'm in the kitchen sanding and painting the cupboards and I look down at him from a short ladder with a simultaneously dazed and horrified expression. My heart thumps and struggles to find its natural rhythm as I realize it's an incensed neighbour who speaks broken English. Red Eyes and his clan are snacking on his vineyard in the warm sun and I gather he wants us to do something about this. His arms waive wildly and spit flies from his frantic stutters. Andrew stands stunned and speechless as I translate the bits and pieces of what I can understand. We are the neighbours to the west of the family with the unruly goats and his vineyard is the property to the east. There are no fences between our properties (as of yet), so the goats wander to both of our properties and eat grapes, or flowers, or whatever they choose, whenever they choose. He explains as best he can and suggests that maybe we can hatch a plan to keep them off of both of our properties. I'm shocked that he already knows about us and about our dismay at Red Eyes crossing onto our property whenever he feels like a snack. News here travels fast.

"Meet our neighbour I say to Andrew," I begin with a sideways grin.

Shortly after our move across Canada, we sat at the bank across from a middle-aged woman with tight curls and thin, gold-rimmed glasses. She peered over them and raised her eyebrows when speaking to us. I felt like an insect under a microscope.

"So you have no plans to find jobs in the next while and you want to start a business with no income?" she summarized, allowing the wrinkles in her forehead to join together in long, stern lines.

"Yes, that about wraps it up," I said enthusiastically, handing her our business plan to start a small wine tour company.

Andrew sat up straight next to me and explained, "This is just the beginning of the wine industry here. There is so much room for growth."

She acknowledged his attempt to convince her that our business idea is a great one, but it clearly wasn't enough. We must have failed miserably at our presentation, because we were denied a business loan almost immediately.

We use what remains available on our credit card limits and scrape the bottom of our savings instead. This funding is much needed for down payments for a passenger van, insurance, getting a website up and running and for flyers and business cards. The rest will have to wait.

Andrew and I visited Niagara to see friends and family at Christmas just a few years prior. We were both working in the restaurant business in Calgary and wanted to further our wine knowledge and familiarize ourselves with the wine scene in Niagara. It was pre-Siri and pre-cellphones and we drove in circles. The signage for the "wine route" and the wineries wasn't all that great back then and we became beyond frustrated when trying to find our way from winery to winery ate up half our day. We had travelled in Italy for a wonderful three-week honeymoon and noticed many services that toured English-speaking guests around the wineries and wondered why the same service wasn't readily available in Niagara.

We arrived back in Calgary with that idea freshly etched in our minds and swiftly researched the possibility of launching a wine touring service in Niagara. After countless Google searches and phone calls, we decided that this could be a great business idea and a way for us to earn a paycheque while we plotted the next step regarding planting the vines and bringing the vineyard to fruition... literally. We came up with the brand Crush on Niagara Wine Tours and schooled ourselves in the ins and outs of licencing requirements, permits and launching a new business in Ontario. We asked each other, how hard could it be?

Gold Rims at the bank treated us like we hadn't the

slightest idea of what we were getting ourselves into and, I suppose, looking back, although we researched, plotted and planned, we may have appeared a bit over zealous and overly confident, but thank goodness for youth and naiveté, because we refused to let dubious bankers stop us in our tracks.

3. Calgary

I didn't plan to leave home, leave Niagara, all those years ago. In fact, I wound up in Calgary on a bit of a whim. When I was 22, I was living in Niagara and working in a restaurant. A friend and two other girls were driving to Banff, Alberta and they were looking for someone to ride along with them to share in the driving and travel costs. I leapt at the opportunity. I had a car of my own and I was certainly ready to take the big leap and try something new.

It was decided that we would leave that spring and stay in Banff for the summer if we could find jobs easily. One of the girls had a sister who lived there and knew the area well. Despite my thirst for adventure, knowing this was a great comfort to me. I thought to myself, what's the worst that could happen if I go? I knew it couldn't be that bad. I also knew that for me, at that time, staying in Niagara just wasn't soul-fulfilling enough. It was time for me to stretch my wings.

I packed up my car and my belongings at the house I shared with my sister Cathy and best friend Tammy and off I went. Saying goodbye to my friends and my family was heartbreaking. I had the full support of some people, while others held their tongues and saved their whispers of doubt until my back was turned. But I knew this was the right thing to do at this time in my life.

It was early morning just before sunset that I began my journey. I felt both exhilaration and apprehension when I got into my steel blue Nissan and drove away from all the comforts I had ever known. There it was in the rear view mirror getting smaller and smaller, the place where I grew up, the people I had come to know, the restaurants I had worked in, the boys I had once loved and the ones I thought I had. The first hour of driving was the hardest, the memories flooding in, the good times and the bad. Saying goodbye to people I loved and people I didn't want to leave. Waves of uncertainty and anxiety hit as I headed west along the Q.E.W. and Highway 400, reaching the 8000-kilometre TransCanada Highway, five hours later, near Sudbury, but I kept driving.

My car was packed so tightly, I could barely see out

of the rear windows. I had only one passenger, one of the girls that I had recently been introduced to. We were paired up in one car and the other two girls were in another. We tried to stay in tandem but without cellphones, this proved to be impossible. Our only semi-successful attempts at communication on the road were hand signals to each other, an arm waving wildly out of a window or rapid honking followed by a finger point to pull over.

Our cars would start out one behind the other at the beginning of each day, and then we would lose sight of each other on last minute gas fill-ups and on busy highways. We were young, free and full of joy and hope. We would chase each other's car, and try and have loud conversations at each other across lanes, from open windows. A friendly game of cat and mouse would ensue almost every afternoon. That ride to the mountains is a memory that I will always cherish. It was a newfound freedom leading me down a path that would change my life forever.

The drive out to Banff took five long days. We were hot, then cold, lost, then lost for words. Going to a place I had never been, with people I barely knew, with nowhere to stay seemed absolutely preposterous to me at times. At other moments, it seemed daring and exciting, which was exactly what I craved. The drive was equally tiring and exhilarating. My passenger and I laughed, sang and fed our bodies a plethora of junk food and oversized greasy lunches from too many 7-11's and sketchy looking roadside diners to count. We were exhausted from the drive at the end of most days, checking into inexpensive hotels, always seemingly missing letters from their signs and always somehow equipped with the lumpiest of mattresses.

The last day of driving was the longest. The green highway signs promised Banff to us over and over again, 600 kilometers, then another 400 kilometers. We pressed on and it seemed like time was standing still. We craved more stops, more time to stretch our legs. We wanted to be done with the driving. Then there they were, ominous mountains either side of us with nothing but my future in between, begging me forward. We finally arrived in the quaint little town, late in the day, colourful buildings lining the streets like a row of Christmas presents waiting to be opened. I felt elated.

I made many new friends during the first few weeks of settling into Banff. We acquainted ourselves with all the nightclubs; a seemingly new group of people arrived almost daily and we all often found ourselves stumbling home with the sunrise. We climbed several mountains and hiked around deep blue lakes that looked like ink pods. Banff had the freshest air I have ever tasted; intoxicatingly crisp, like biting into a gleaming apple where the fine sweet mist sprays refreshing dew. I had daily run-ins with elk, deer and even bears. For the first time in my life, I began to connect with nature in a way that I hadn't had the opportunity to before.

We often went for long hikes through the remote, rugged mountains, starting in the morning, coffee in hand. We would take food and drink breaks every few hours, laughing together while putting our faces toward the sun. Each of us holding our hands above our eyebrows and squinting at what lay before us. From there on the crest of the mountains, the views are breathtaking, seeming to go on forever.

An inexperienced hiker - and not in the best of shape, as a result of the long nights of partying, I often lagged behind the others when we started out again. One day I lost sight of my friends up ahead.

It was mid-afternoon on a sunny day on the perfectly still mountain when I looked up and they were gone. I looked ahead, looked further, straining my eyes hoping they were there somewhere blending in with the cold grey stone and blue sky. The clouds rolled on thick and white like ships passing by me lost at sea. I saw nothing. I called for them over and over, but the only reply was my own echo. Had I gone off the trail? Had they? How long ago had I lost them?

I raced downward into the thick brush calling and calling them. I became frantic. My heart pounded. I started sobbing. My legs shook. But there was nobody and nothing around. No animals, no people, not a bird in the sky. There were no sounds. The silence felt oppressive. I scaled back up the side of the mountain towards the path that had let me down. I sat on a cold, grey slab, shivering, alone and defeated. I had spent years trying to find my path in life, find out who I was and who I wanted to be, and now here I was alone at the top of a mountain drowning in nothing but myself.

I sat on that rock for what felt like half a day, likely only two hours or so, then finally, incredibly, I heard a faint sound. It was my friends calling to me, peering into the distance, I saw their miniature figures waving. I ran on shaky legs, more relieved than I'd ever been to see those familiar faces. They laughed, smiled and shook their heads at me. I dried my tears and tried to pretend I was okay, and not someone who completely fell apart only moments ago, thinking I would never see anyone again.

We descended down the rocky side of the mountain onto green grass. I know I am lucky to be alive. Had I continued back into the thickness of the trees and rocks, I may have never found my way out. Had they not found me, I am not sure to this day if I would have ever left Banff or that mountain.

Living in the mountains, I met the most interesting people. Ballerinas who danced like swaying willow trees in gentle breezes, friendly and courteous Australians beginning their treks across Canada, people who searched for quiet and people who were simply searching for themselves. Maybe it was how welcoming the fresh new faces of Banff were, or my newfound sense

4. Blood Sweat and Years

"Do you see it?" I ask Andrew, "What it might be one day?" We are in the upstairs bathroom. It has the best bird's eye view of the entire back ten acres. He leans on the windowsill and stares out at the rectangular property, out at the dead grass and trees and old concord vines that no longer bear fruit.

"It's going to be so incredible one day," he says.

I once read in a magazine that it takes ten years to completely restore a property from start to finish. I relay this to Andrew and realize immediately that my timing is off. Ten years seems like an impossible length of time right now.

He looks at me and sighs, "Let's just take this one day at a time for now."

"Good thinking," I muster.

We have to continue to believe that anything is possible.

I try to make myself take things day by day instead of looking at everything all at once. I find it too overwhelming to be continually adding things to my checklist. It's like looking up a steep trail to a mountain and feeling defeated before you've taken the first step. And I know how hazardous mountains can be. It's just dangerous anxiety territory and I don't have time for it now. We are close to moving into the house now and fall is right around the corner.

As well as all the work and the financial worries, I miss Calgary and our dear friends that we have left behind. It's a loss that hits me in waves; faces of people I'm used to seeing every day are no longer around. I miss their voices and running into them by chance at our favourite local market or coffee shop. I long for the smells of the city and the hum of the traffic. I miss the convenience of walking out my door and walking down 17th Avenue with its hip restaurants and coffee shops. I miss the feeling of Stephen Avenue at night, watching people walk from bars and restaurants their breath and cigarette smoke trailing behind them in the crisp air. I miss walking into my favourite wine bars and knowing the bartender and the entire wait staff.

I lived in Banff for two years, but then felt boxed in, needing to feel the energy of a big city, so I moved to Calgary. I explored the quaint little hamlets of downtown where there were chic bars and eateries, each neighbourhood with its distinct and charming personality. I became immediately smitten with the shops in Kensington and the vibrancy of 17th Avenue. The grandeur of Mount Royal and various other well-established and coveted neighbourhoods had me swooning. I found the best place to eat a homemade fresh tuna sandwich on 4th Avenue and the place that served the best coffee and the fluffiest pile of pancakes on Sunday mornings. Andrew and I loved walking by the large two-story brick houses on warm sunny afternoons wondering who filled them. On weekend mornings, we'd have pajama parties when friends would pitch up early and fill our small house with laughter, the smell of coffee and toasted bagels. We would eat omelettes and warm baked banana bread and catch up on our week, then around noon everyone would disperse for naps. Many of us had homes close to downtown so we were always within a short walk of each other's houses.

Calgary is where Andrew and I met. We worked together in a restaurant, aptly enough. I walked into a restaurant just off of Macleod Trail and hit it off with the manager almost immediately.

"What brings you to Calgary?" he asked.

I replied with a smile, "Fate, I guess?"

At that moment I didn't realize quite how fateful it really was. He hired me on the spot. I needed the job desperately. I was staying at a hotel nearby, having only just arrived a few days beforehand. I didn't know a single person in the city and had just enough cash for a first and last months' deposit on an apartment when I managed to find one.

Andrew was away my first week of working there, but we met at a dinner party at our manager's house. He walked into the house and walked straight towards me. He told me I was beautiful and sat next to me on the sofa. He brought me too many White Russians to count and made me laugh all night.

Andrew grew up suburban Calgary in a traditional middle-income household. Like most kids growing up in the 70's, riding his bike and skateboard offered him a freedom like no other. His friends would meet at the local corner store and buy dime candy and drink pops until their bellies swelled. His parents worked hard to afford him a private education at Strathcona Tweedsmuir School. They weren't like some of the uber wealthy families there who handed fancy cars to their children for birthday gifts or offered exorbitant holidays to faraway places for good grades. Andrew's parents made sizable sacrifices to send him and his two brothers to this renowned school where good marks and perfect attendance were mandatory.

He was influenced by his own parents, but also by those of his friends. He saw the wealth accumulated by families who worked for themselves - and the good lives they built for themselves. It was the affluent entrepreneurs who started from very humble beginnings and built businesses from nothing that would influence him later in life. He wasn't much interested in the kind of extreme wealth that wore brand names and flashed fancy labels. He, very much like me, was interested in building something from scratch, something of one's own, something gratifying.

Andrew was a big dreamer, and again much like me, he found school couldn't cultivate his curiosity. Sitting at a desk in a rigid blazer and tie, he tried to stay focussed on books, work and words on the chalkboard, but he was drawn by the travel books he saw in the library, with their blue skies, beaches and faraway hotels populated by people with fresh tans and broad smiles.

He, too, longed for a livelihood where he felt like he could flourish.

It would be a TV mini-series called *Hotel* that he watched in his early teens that would further provoke his interest in travel and entrepreneurship. He loved the idea of having a career that also offered a varied lifestyle and a livelihood where he would be his own boss and not have to answer to one.

So, after high school, Andrew began a course in

restaurant and hotel management and took a job as a banquet waiter at the renowned Palliser Hotel in bustling downtown Calgary. It was definitely an entry-level position but it paid the bills while he dabbled in university and moved into his first apartment. It certainly wasn't his dream job or his forever job but it was a good "for now" job and a stepping-stone toward greater things to come.

It was at the Palliser that Andrew developed a love for food and wine and he went on to work in some of the most revered restaurants in the city as a server. His love for wine grew while working for Richard Harvey at Metro Vino, a charming wine shop in downtown Calgary. Andrew then went on to earn his Sommelier accreditation. Every success story has humble beginnings and his was no different. Passion takes growth and growth takes passion. Andrew was and still is wildly inquisitive. His passion and path to success came from his humility and his willingness to learn and ask questions.

"I can't fucking believe it!"

I shout wildly with hand gestures to match. I'm seething. Red Eyes has come over in the early morning and eaten every planted flower in our front yard. The stems sit headless swaying in the breeze like green liquorice sticks.

"That's it."

I storm next door. Andrew stares after me, wondering how this is going to end. Our neighbours to the right are strange people. Their house always seems to be dark with the curtains drawn and they are pale and thin like farm country vampires. We've seen them only a handful of times since moving in. They completely freak me out, so my confronting them on their turf is a bold move.

Bang, bang, bang, I hammer on their front door. Bang, bang, bang, I continue hammering. I stand my ground when it whips open.

"Your bloody goat ate all my flowers on our front lawn. Can you not keep them fenced in or something?"

The male farm country vampire is obviously mortified.

His hair is greasy, black and slicked onto his forehead like an ill-fitting toupée. His mouth opens, but no words come out and he looks as though he might faint.

"I'm so sorry," he finally mutters, "I'll take care of it right away."

His lips don't touch when he speaks, the bottom one shakes slight and I find it unnerving.

"Fine," I say, "That was about $400-worth of flowers."

He walks away backwards until his long fingers slip away from the edge and it latches closed.

It's the day after the headless flowers incident and we take a break from our monumental "To Do" list to walk through the back ten, hatch the plan for cutting down the old orchard trees and work out what it is going to be required to clear the field and get it ready for the replants down the road. Andrew pulls out a large picnic blanket and whips it open with a quick snap. He lays it on a patch of grass on the back ten and we dine on bread, cheese and a bottle of local wine. The warm red liquid is smooth as silk and its gleam is intoxicating when caught by the sun. The Cabernet offers a temporary respite and a taste of civilized normalcy.

"One day we are going to bottle an amazing wine from this farm," Andrew says delightedly.

He holds his wine glass by the bottom of the stem and tips it up toward the sun admiring its inky red through the rays of light. I smirk and nod, thinking that it's going to take forever to get to that stage but holding it in the realm of possibilities. I lie back, hands behind my head, watching the thick clouds soar by on the sea of blue. I want to lie on this blanket for the rest of the day, a life raft saving me from the blood, sweat and years of hard work that is to come.

"Let's get back at it," Andrew says as he lifts the blanket so high I roll back onto earth, back to reality.

We trek back to the house feeling relaxed and satiated when I feel Andrew squeezing my arm as we get closer to the main house.

"Oh my God," he barely whispers as he looks ahead, wide-eyed.

My pace quickens.

"No, no, no!"

I can't believe what I am seeing. We tiptoe forward and peer through the neighbour's fence and there it is, a furry, matted, white bloodstained mess with flies swirling around. The neighbours have killed Red Eyes and his head lies there, a gross and ghastly reminder of how cruelty and ugliness can crop up even in the most wondrous and beautiful places. I am traumatized by the sight of him and wish I could hit the rewind button and pet his matted fur, lay a flat of his favourite flowers in front of him and tell him to munch away to his heart's content. Had I known that when the neighbour said "I'll take care of it," that he meant he was going to kill Red Eyes, I would have certainly found him a good and loving home. Rest in peace, Red Eyes. May you be romping happily through the greenest of grasses and savouring the most brightly coloured flowers. I am so sorry.

5. Progress

I run my hands down the cold front of the stainless steel fridge approvingly and I am thrilled at how far the kitchen has come. The cupboards are painted a fresh coat of dove white. The thick cream paint offers a reprieve from the yellow, dirt-stained cabinetry and gives the kitchen an immediate new lease of life. Andrew and my brother-in-law Kevin make me the kitchen island that the space begs for. We buy inexpensive, knotted, turned banisters and screw them together with some stained pine planks. The end result is crooked and leans down a little on the right side, but it will do the job for now. I priced out several other options all over a thousand dollars. This one comes in under two hundred, which makes me love it even more. It sits proudly in the centre of the new space and I smile to myself thinking how far this place has come already.

The air is crisp and fresh and it smells like fall through the open windows. Apples dot the trees behind our house and I fantasize about baking apple pies from scratch and filling my house with the scent of wafting cinnamon, but there's no time for that luxury. We are so close to moving in. We need to get the handles on the cupboard doors and drawers and sand the bottom of the doors to accommodate the new hardwood floors. The kitchen's front and back doors only open halfway because with the new hardwood the floor is too high at points. Every time someone comes through, the doors swing open and abruptly halt, leaving a fresh new scratch, a constant reminder of our oversight. On top of these unsightly under door scrapes, the gap is just big enough to let in mice and other unwanted rodents, one of which ran straight into the kitchen and up Andrew's pant leg early one morning! So fixing the doors has been added to our seemingly never-ending list.

I can't wait to hammer in nails on the walls and hang fresh memories and watch the house become our own. Andrew wants to get rid of all the half empty paint cans. There seems to be a paint tray in every room with a paintbrush wrapped in cellophane and begging to be dipped in a new coat of fresh possibilities and touching up spots that we've missed or drips that we didn't catch.

Each time one of us walks by, we grab the brush and touch up a spot or two that needs attention. This has gone on for weeks. But the walls look as good as they are going to for now. It's time to call the painting quits. We have to set up the office and get the touring business up and running.

Getting Internet to this farmhouse presents another challenge. I have the website ready after countless hours of revisions. The conversation with the tech person on the other end is beyond frustrating. He is barking orders at me and I feel like I am trying to land an airplane in the dark. I am typing in codes and a skull and crossbones keeps appearing on my screen. It's after midnight and after a long workday, this is the last thing I want to be dealing with.

Getting licencing for our vehicles is yet another story. We set up the office and begin printing up our flyers for our wine tour brand. It isn't until we go to the regional police that we realize a licence in the area we want to tour in doesn't even exist. So we simply call the mayor of Beamsville, sit him at our coffee table and tell him that he needs to create one. We have gone too far down the rabbit hole now (and desperately need an income) and I won't take no for an answer from anyone, not even some of the naysayers who come by our house to welcome us to the neighbourhood, but say things like, "Wow, you guys are young. Are you sure you know what you are doing?"

It's become a running joke amongst friends that our new fridge has nothing in it. We have been living lean and I'm now fretting over every penny. Frantically checking our bank account daily has become a nasty habit and I'm really concerned now that we won't make it in time. We have exactly one car payment left in our bank account. Other than that I have no idea how we are going to buy groceries or pay any utility bills.

It's a Friday night and my girlfriend invites us to a fundraiser because she has extra tickets. Our social life is non-existent except for the brief chitchats with the tradespeople that frequent our house. We arrive at the community hall where people smile and hand us foamy beers in plastic cups. I feel vaguely normal and am glad to be out at night amongst a crowd of people that I know. The raffle tickets cost $5 each and Andrew pulls out a $10 bill to contribute to the cause of building wells in Africa and I shoot him a look of concern. He knows exactly what I'm thinking and I'm embarrassed that I may have made it obvious to everyone around me, too. Andrew gives me the look - eyelids lowered to half-mast and a tilt of the head, which means, "It's ok, don't worry about it, relax and have fun." I take a deep breath and acknowledge that I understand what he's said, without saying it, and we both grin while sipping on our drinks. I have two rectangular shaped tickets in my hand with seven-digit numbers and I fidget with them in my pocket throughout the evening. One side is rough from where it was torn from its partner and I can't stop running my thumb along its edge. It's 10:00 p.m. and my girlfriend's brother-in-law announces the draw. One of the first prizes is a large straw basket filled with pasta, sauces and all sorts of Italian foods.

"God, we need that," I whisper to Andrew as the numbers are read out. I mentally count off each number as each one is called aloud and I squeal with delight as I fire my hand into the air. That tiny ticket number matches the food basket!

"Yesss!" I shout into the tight crowd. The weighty basket is handed to me while everyone claps and smiles, my closest friends beaming. There are passes to a local spa - something I have no time for these days, tickets to live outdoor performances and several other services offered by accountants, book keepers and so on. But I am only interested in the food baskets - anything to fill our fridge and sustain our bodies over the next month or so. The next basket that I want is filled with jams and breads and cookies. It's breakfast for a month.

I say to Andrew confidently, "This one is ours,"

Sure enough I hear the numbers called and they match my ticket. I'm embarrassingly elated.

I scream, "We have food! We have food!"

People around me laugh, point and nod their heads. Many of them shoot a thumbs up.

We arrive home no longer feeling defeated. Seeing food in the kitchen and the fridge without vacant shelves fills me with sheer delight and promise for the future.

"This is going to happen," I keep reminding myself. "We are going to make all of this happen."

The website for the tour company has gone live and I am in the office waiting by the shiny new phone. Our business officially exists, and I go through a mental checklist of all the things that had to happen to bring us to this point. Andrew orders shirts and name tags for all the staff. All two of us. Andrew and I are the only two employees of the company for now.

"Here we go," Andrew says handing me my Crush on Niagara nametag. "Now the real work starts."

I feel my heart and stomach flutter. Andrew is licenced and ready to go, as is our one and only touring vehicle, a 15-passenger van. We have connected with all of the wineries and local bed and breakfasts. We have even knocked on doors telling people about who we are and what we want to accomplish. Now we just pray, we ignore all of the naysayers, we believe in ourselves and hope that it all turns out the way we intend it to.

My sharpened pencil lies waiting by the newly acquired

reservation book. It is a thick bound green book with a black spine and it's so fresh that when I open it, it creaks. It smells of new paper and a new beginning.

It takes only a few days for that first phone call to come in. My hand shakes when I pick up the phone.

"Crush on Niagara," I answer almost fearfully.

It's a woman looking to book a tour for four people on one of our daily touring programs. My voice is shaky and I look at Andrew with wide eyes. His hand covers his mouth in disbelief. I book our first guests, hang up the phone, and cry happy tears.

Shooting my arm straight into the air, fist to the sky, I shout with glee to Andrew, "It's working! It's working!"

It was only a short time ago that, when visiting a winery and explaining our business model, a young woman stood before us and said, "You know, guys, people have tried this before."

Her words were like a dagger deflating an unsuspecting

balloon. We knew instinctively that she was wrong and planned to prove it or go broke trying. This first booking tells us that we are on our way.

Our income trickles in as the tour company begins gaining momentum. We hire one more tour guide, then two, then office help a few mornings a week.

"Slow and steady," Andrew says, "One step at a time."

We keep taking steps.

The vineyard still needs to be completely cleared and the vines planted. The house is still very sparsely furnished, with the exception of a few beds picked up from a quaint little antique shop down the way. The living room and den furniture have to wait, as do the rickety, drafty, old windows with chipping paint. We sleep by the wood-burning fireplace on the floor most nights that first winter to stay warm. "Drafty" doesn't come close to describing the house. The winter winds bite through the old siding and the floors are so cold I constantly have foot cramps. The plastic shrink-wrap that we blow-dried against the main floor window frames flaps all night in the icy wind. It's pushed into the room, and then sucked out again like a jellyfish pulled by the tide. There is so much more to do, always so much more to do to make this place feel like a comfortable and finished home. But the house will have to wait for now. We are out of money and time. We will have to wait to fill in the cracks and the pieces of our dreams that remain.

6. The Third Mortgage

I look out the window at the small bungalow to the west of our house and see what appears to be an old sofa with bits of frayed orange material flapping from its carcass. I see chairs - one, two, three, four of them - strewn on the front lawn. Two lie on their sides with their legs bent and splintered. They are broken... and I wonder if the person that threw them had been broken, too, as they appeared to have been hurled with such indignation.

Andrew looks out the window next to me wondering what's got my attention.

"Is that furniture on the lawn? What the hell, it doesn't look like a garage sale..." He wonders. "Let's go check it out," he says aloud leading me out the door.

We wander out the front door and slip over the intervening forty feet to see what is going on with this recently vacant house. We tiptoe like FBI agents on a sting and I feel like we are doing something that we shouldn't be. All we are missing is a dark sky, a flashlight and balaclavas and we are two criminals ripe for the picking. Two of the basement windows are smashed and the jagged pieces hang from aluminum frames. They remind me of shark's teeth, an unfavourable start to this exploration. They are the house's ominous warning, two big serrated black holes that seem to form one large mouth at the bottom with the dirty windows above looming over us. "Stay away or I will bite you..."

The rickety back door is ajar and it's dark and damp once we push it open.

"Hello! Hellooooo?" I call, but there's no reply.

I'm in front of Andrew and he's nudging me forward until I swipe his arm away.

"Go, go!" he says.

"You go!" I mouth, and we switch spots.

His head is sticking forward out of his shirt collar and

he is cupping his ear with his hand straining to hear.

He shakes his head and mouths, "Nothing?"

I shake my head. We take this silence as an invitation to sneak inside. The smell hits us immediately. The acrid stench of urine swamps us like we've just been hit by a huge wave. The odour is so sharp it makes our eyes water, so we cover our noses with our forearms and let our shirtsleeves spare our noses and lungs from the worst of it. We inch in with our backs to the wall and slide stealthily around the used needles. There are large, white circles stripped from the hardwood where animals have peed and claw marks ripped into the forgotten floors. It's the blood that really gets to me. Faint sprays of red on the kitchen wall and another bloody trail on one of the bedroom walls.

It is hard to believe that the toilet could have once been white. It's the stuff of nightmares, a black gunk coats the bowl and toilet seat, creeping onto the toilet lid. It looks like someone put black licorice and water in a blender and flung it at the unsuspecting fixture and I immediately think back to the movie *Trainspotting.*

We look at each other like two people underwater trying to communicate. Eyes wide, holding our breath. Not saying, "What the hell happened here?" Knowing we're better off not knowing.

The walls are dripping with a tobacco stain-yellow sticky substance that hangs from the filthy popcorn ceilings about a third of the way down. There are two bedrooms downstairs, each with torn window screens and matching hot pink window frames. Upstairs the half story has two bedrooms and one large, dark bathroom covered in a sickly orange faux wood.

There's a dirty blue notice on one of the front windows and Andrew goes outside and pulls it from its tape. It's a Notice of Foreclosure from the local bank. I feel like I do when I see a new pair of shoes that I really, really want but I don't want the shopper next to me to notice that I'm smitten. My pulse quickens, but I pretend to play it cool underneath the guise of indifference. I feel that way here in this abandoned, neglected house. I get goosebumps and think to myself, "I wish I could get my hands on this place and make it pretty," but I say nothing. I will keep it open in the realm of possibilities.

Less than a week later, we hear the loud deep roar of a motorcycle, the kind that vibrates in your stomach if it's close. A succession of obnoxious "vroom, vrooooom" noises follow, coming in short, loud, violent bursts. We peek out of the window and see several men on motorcycles pulling up outside the neglected bungalow. The largest of the men arrives first and hops off his bike. He is a mountain of a man with a grey, wispy beard and a shirt that is missing its sleeves. He eagerly shakes the hand of the man in the suit and tie waiting outside and we strain to hear their conversation.

"Oh my God," Andrew says, "Are they buying the house?"

Andrew moves closer to the window, then creaks open the door. I don't hear anything.

"Can you hear anything?" I ask

"Shhhh..." he begs.

His eyes are closed and scrunched and his concentration is unmistakeable. He reminds me of a doctor listing to a heartbeat through a stethoscope.

"They are talking about the cost of the house and the size of the property."

The breeze carries some of their conversation. Andrew can only catch bits and pieces. They are too far away.

Our backyard isn't fenced and we aren't certain of the property line at all. It is more like two houses on one shared property, but the bungalow is a small square carved out of the long and narrow ten-acre plot. All we know is that all that stands between that house and our vineyard are three graceful willow trees. Looking at our potentially prospective neighbours, we realize that is nowhere nearly close to enough. Our bedroom window overlooks their undersized backyard and we have immediate visions of all night parties, bonfires and bikes roaring in at all hours. Worst-case scenarios play on a frantic loop in my mind and I have visions of nothing but pandemonium and empty beer bottles.

Apparently Andrew sees the same.

"We have to buy the house," Andrew says and I know he is right.

"How are we going to do that?" I question, mentally searching through all the financial avenues that we have already exhausted.

We are showing a little income with the tour company moving in the right direction now, but not nearly enough for another sizeable loan.

We call the bank the next day and the person we speak to can hardly contain her sarcasm.

"You want to buy the house next door?" she scoffs.

I can hear the crackle of her lips as her mouth makes a dry upward smile and I am sure she's trying hard to contain her disbelief. My palms are sweaty and I dig deep. Hope here is an undeniable necessity.

It isn't long before we find the right person at the bank willing to unload this distressed asset. We leap at the chance to assume the mortgage and come up with a very creative way to produce the required deposit within twenty-four hours. We have two credit cards left that haven't already been maxed out building our new life. These rectangular little plastic jewels are the difference between living in peace or potentially living in chaos. There's no turning back now, anyway, and when you have nothing to lose you find the gumption to dig that hole just a tiny bit deeper. You just push aside the gnawing fear and uncertainty and go for it.

We get the keys to the house soon afterwards. We've had enough time to plan on turning it into a guesthouse and short-term vacation rental. I've got a new clipboard and a new to do list. We roll up our sleeves and go to work immediately. Walls come down, bathrooms are gutted and floors are pulled up within days. Another step, albeit an unexpected one, taken.

7. Haul and Burn

Andrew and his dad Harvey continue clearing the old orchard that winter into spring. There are still old concord vines and large stumps that cling to the ground, their roots grabbing at the earth like long willowy fingers. I can still hear the roar of the chainsaws and the hacking and sharpening of the axes. Hollow old cherry trees fall to the ground with mumbled thumps while other old pear trees hang on to their rotten roots and endeavour to remain vertical. This is where "Old Betsy" comes in handy. Betsy is our mint green farm truck, a relic with sizable rust holes in the floorboards and deeply ripped seats that reveal the benches' stuffing to be the colour of sponge toffee. Betsy and a long shiny workhorse of a chain pull at rotted stumps and trees for months.

For months, the sounds of the old truck spinning her wheels and the chainsaw humming are always in the distance. Andrew and Harvey wrap rusty chains around the dead tree stumps that remain. Next, one of them jumps into the front of the truck while the other directs the driver from the rear-view mirror.

"Pull, pull! Okay, it's out!" I hear one of them call.

Then that stump gets thrown into the back of Betsy and this continues until the flat bed is full. All of the pulled stumps, thick, dead brush and old concord vines get dumped onto a large burn pile until the pile is ready to burn. It's a process they now have down to a science. It's hard, sweaty labour. There are acres and acres of this to conquer, so they do a quarter acre at a time.

There are constant burn piles of wood and gnarly, mangled old concord vines being lit in rotation. Andrew and Harvey haul and burn and haul and burn some more.

It's late afternoon when I slip onto the sofa for a little respite, but wake up to see a man on our front lawn hopping from foot to foot like he is stepping over hot coals. I wipe the grogginess from my eyes and make my way to the front door where I see more people standing and staring. Now I count one... two... three trucks, plus people standing on the side of the road in front of our house, frowning disconcertingly.

So tired that I'm struggling to focus, I call Andrew's cellphone. I hear muffled noises and he is panting heavily. I realize he is running.

"Call 911!" he says. "Call 911"!

I run to the back window and there I see it. The entire back ten is in flames and there is smoke as far as the eye can see.

"Fire, police or ambulance?" the dispatch asks.

I scream, "We need a fire truck, h... h... huge fire," I struggle to find the words.

I force the address out of my frightened head, slip on my open toe shoes and run out the door.

I see people everywhere stamping out little piles of flames that bounce from one bunch of brush to the next.

It's fire leapfrog and everyone is pitching in to help. People I have never seen are stamping and jumping, a constant stream of people running onto the property like there's a finish line somewhere at the end that I can't see. There must be twenty people now, each stamping out their own pile of flames. The volunteer firetrucks show up in a row, one shiny red truck after the other, sirens blaring, and now everyone in the neighbourhood knows that "them city kids" with the black 4 x 4 set their backyard on fire. I'm embarrassed and humiliated.

I stomp my foot on mini fire piles one after the other until two hours have gone by and the fires are finally out. We are all parched and exhausted. I have soot all over my jeans and it's crawled up my ankles like a dark shadow of an unwanted reminder that we are in over our heads. My Birkenstocks are smoking, literally, my feet are blackened and Andrew's face is soot-covered with only tiny streaks of his usually pale complexion peeking through.

We survey the damage. Andrew confesses to me that when he was standing in the middle of the vineyard, surrounded by smoke, he was struggling for air.

"Jesus," I say, horrified, "You could have died."

He continues, "I held my breath and ran as fast as I could until I could see people." He pauses, shaken. "I had to find my way out of the smoke," he says to me, wide-eyed.

I just count my lucky stars that he had the wherewithal to be so resourceful and that he kept a cool head and made it out unscathed, but I know his ego is bruised. The property is black ash, all of it, still smoking like a desolate and barren wasteland of sadness.

The people that selflessly helped put out the fire introduce themselves one by one as they say "good luck" and depart. Looking back, they must have pitied us and at the same time thought we were way out of our element, and well, we were. It turns out that the fresh burn pile from that morning wasn't put out completely and the wind picked up and ran with it when we turned our backs and got on with our day. It was a rookie mistake. A near deadly one.

We are learning. Every day we learn something new. The plumbers' words ring clear to me now, when I questioned what exactly he meant by our cistern needing water shortly after acquiring the property, I naively asked him, "What do you mean? Where do we get our water from?" He laughed, shook his head, scratching at the back of his neck and said, "You're livin' on a farm now, girl."

I look out the window and see Harvey flapping his arms and waving his hat in a strange sideways figure of eight motion and then start running, while Andrew drives Betsy at breakneck speed in the same direction.

"What in the hell?" I wonder. "What now?"

I lift the squeaky old wooden window frame to get a better look and hear Andrew scream, "Run, Dad, run!"

Andrew hangs out the driver's side of Betsy, one hand on the wheel and one on the door, the truck bumping over the ground. He reminds me of the kids we saw in Africa dangling off the side of the trains, but he looks far less content than many of those kids did. Harvey bolts alongside.

Harvey makes it to the house breathless with Andrew

running after, the truck abandoned behind him.

It turns out that they turned over a log with a wasps' nest and Harvey got the worst of the irritated insects' wrath. He suffered nearly a dozen stings, but is okay.

"Good Lord," I think to myself.

Every day in the country is so unpredictable and so full of surprises, both good and bad.

I am still at the helm of the tour company, on the phone, figuring things out, flying by the seat of my pants, learning something new each and every day. We've bitten off a lot. Some days I laugh a lot and bask in self-accomplishment and other days I feel like I've run flat out for ten hours and I'm heading for a cliff and I can't stop so I just allow myself to go right over. I imagine that my adrenal glands look like shrivelled, dried up olives and I just want to hit pause and rest, but I can't. I start having panic attacks almost daily. For a while it is debilitating and the fear grips me and stops me in my tracks. Self-doubt and worry consume me and the uneasiness won't let me go. It's like I'm on the shore of a massive looming body of water and a tsunami is coming toward me. It's so high and powerful and imminent that all I can do is cower. I can see it over my head, ominous and impending, putting me in a shadow, blocking out the daylight, darkening the sun. I keep running from it, running from everything, running from myself. I keep moving and doing. I watch the tidal wave emerge and come toward me, closer and closer, waiting for it to swallow me, the fear and the panic and the self-doubt and the jobs we've given up and this whole overwhelming, monumental task of starting from scratch and picking up our lives and moving across Canada.

Finally I let it in. I allow myself to breathe it all in, the entire massive project, and I tell myself it's okay. It's okay. It's all okay. It's okay to be scared and to be unsure. It's okay to have taken all of these risks and to be afraid. I wrap my arms around the fear instead of running from it. I allow myself to inhale and look for the sun and the wall of anxiety lands at my feet, freeing me from the darkness. It retreats and I make peace with it all.

We start renovating the derelict home next door in

the spring and I stage the rooms at our own house with comfy beds, pretty pictures and the promise of a vineyard retreat... because we've already begun taking guesthouse bookings for August and the actual rooms aren't even ready yet. Looking back, it is beyond crazy to push ourselves like this. We are hammering in nails and whipping fresh duvets out of plastic bags just minutes before the first guests arrive.

Locals stop by uninvited, some with kind gestures and wishes of good luck, others laden with leering doubt. One woman actually says with undisguised disapproval, "Oh my God, it's just the two of you, you're so young," while another says condescendingly, "You guys can't really be doing this." I lose my battle with politeness and bark back, "We will make it happen." But my confidence is taking a hit. I'm tired of this uphill battle and certainly tired of being judged by my age. Just because we are young, it doesn't mean we are incapable. "The house just needed a little spit and polish," I said in the beginning, but apparently, so did we. We are really flying by the seat of our pants but no one needs to be privy to that information. We've done so much and yet we're still not there.

Andrew calls me outside to the front lawn. We have six beautiful cherry trees that stand guard over the property in two tidy rows. The luscious pink and white perfumed blossoms seem to have opened overnight.

"Stand here," he instructs, and I duck beneath the tree and tuck in next to its trunk in the shade.

He shakes the tree above me and the sweetly scented petals surround me, raining down and reminding me of the striking beauty that is everywhere in wine country.

"I thought this would cheer you up," he says and continues gently shaking the tree.

He has one hand on the trunk, holding out the other like a person trying to catch the frozen flakes of a first snowfall. The pink petals glide down on both of us. It's a moment I will never forget. The smell is intoxicating and it's a brief and welcome moment of elation and escape, all at once.

I go back and fling open all the windows in the house and the guesthouse and let the sweet, creamy fragrance fill our homes, minds and bodies with the promise of a fresh new beginning.

The guesthouse is constantly busy and is a wonderful addition to the tour company bookings. It's a lot of work and in the beginning we do it all ourselves. We welcome guests, change sheets, make breakfasts - Andrew more than I, admittedly, take bookings and grow the business, day by day, week by week, year by year.

We run it at a higher occupancy than we expected for over ten years, with fabulous groups and happy go lucky guests filling the house with wine, laughter and friendship.

8. Growth

"Push, puussssshhhh," I can hear my doctor hurling these encouraging words at me.

I hear visceral grunts and groans filling the sterile delivery room and can hardly believe they are coming from me. Andrew stands next to me with a look of bewilderment and encouragement and he struggles to find the words to soothe me.

"You're okay, you can do this, the baby's coming," he says softly.

I hear the words but I don't respond because I'm too weak and too afraid, so I just brush my hand across his wrist so he knows I can hear him. My doula lends her hand when the contractions come and I squeeze the fragile offering. I fear I will snap the tender bones so I loosen my grip and whip my head from side to side as though trying to hide from pending agony of the next bout of labour pains.

Three hours have passed. The room has been a steady hum of nurses and doctors chatting, the smell of coffee being passed and offered in white paper cups, Andrew coming and going from the room and a steady stream of women giving birth in the room next to me. First screams of pain, then screams and tears of sheer delight. Women come in with swollen life in their bellies and wheel by my open door with tiny bundles in their arms only minutes later. I count three of them and wonder when it will be my turn.

"Oh my God," Andrew says to our doula and to the nurses, "Is this normal?"

He paces around the room doing laps around the bed. He is trying to maintain an unconcerned composure but I can tell he is falling apart.

I feel like I am floating outside of myself. I hear things around me and see them happening but it's through someone else's eyes. I'm a butterfly and my body has gifted me the ability to break from my cocoon of pain and I'm here in limbo waiting for my body to find the strength to keep pushing. I keep reciting the same words to myself over and over in my head. "You have a job to do. You have a job to do..." It keeps the pain from completely consuming me.

Then suddenly the room goes quiet and the hum lifts. I raise my head and see my doctor looking at my doula and the nurse looking back at the both of them. Worry spins a silent web between them and I know it almost instantly. The nurse turns down the monitor that provides me the steady beat of my daughter's heart and I turn immediately and look at her.

"Christina," she says, "The baby has to come out now, her heart rate is dropping."

She is asking too much of me and I am asking too much of my body but I come back into myself anyway, back to the distress and suffering and garner the will to keep pushing. My doula is holding one foot and my doctor the other and I tell myself, "This is it, this is what you've been waiting for, so do it." Two more pushes, the blood rushing to my head.

"Breathe, Christina, breathe. Push, push."

And here she is. This tiny slippery wet human, screaming with wild eyes, hands splayed to the sky.

My doctor holds her above me and says, "Look at her, Christina! Look at this little girl!"

I fade in and out, filled with pure joy and exhaustion. Andrew is crying with his hand over his mouth and he is in awe.

"Oh my God, oh my God," he says in a hushed tone as he fights back tears.

I watch as she is wiped and weighed, and then feel the weight of her warmth on my chest. Out of my body now and on my body, still connected to me, the beating of our hearts matching their rhythms. She is so warm.

Ellery, our first daughter arrives on a Saturday. We brought her home on Valentine's Day. The sweetest gift ever. My water broke the previous day, on a Friday, our busiest day, at 6:00 p.m. when the guesthouse was fully booked. So, naturally, we called the hospital and explained we would be there after our guests had checked in, we had attended to them and got them settled.

Fridays are always rather chaotic. It is a big check in day and people are notoriously late. It isn't uncommon for a 5:00 p.m. check in to show up at 8 and casually explain that they decided to pull off for dinner first. We learn rather quickly that a "self-check in" option is a necessity, and put a key code on the door.

The office for the tour company is in our house. It's temporarily camped out in the furthest spare room. It has served its purpose quite nicely for the first few years. There's a simple pine desk on a trellis in the corner, a silver filing cabinet and some looming black bookshelves that are far too tall, so much so, in fact, that they have scraped the ceiling where they have been squeezed in. This pseudo office served us very well until Daughter Number One graced our lives, but it's becoming increasingly difficult to answer the phone with a baby on one hip, get any rest with the phone ringing at all hours or get the baby on any sort of schedule while trying to run and grow a business just steps from her crib.

We decide that the outbuilding closest to the vineyard is to become our office. The grey concrete has such deep cracks that the slabs are heaving in completely different directions. The barn must have been beautiful when it was originally built but by now, time, decay and neglect have taken over. The once vertical boards lean on one another for support and the rusty nail holes at the tops have dissolved, leaving long, narrow gaps where birds swoop in and out as they're busy building nests.

We survey the damage with our good friend Dan. He owns a little winery up the road and is a third generation farmer who has grown up in the country.

His first recommendation is, "Tear it down and dig a BFH."

He surveys the barely standing barn with one hand on each hip.

He adds, "Wow, you guys really got your hands full

here, eh?"

Dan is a Viking of a man with baseball mitt-sized hands and a broad smile. He sports a farmer's tan and gleaming white teeth that surprise me every time he smiles. Andrew and I look at each other and his slight gesture of his shoulders upward with the tilt of his head to the side signals to me that he has no idea what Dan is talking about either.

I say, "BFH?"

Dan lets out a hearty laugh and says, "You guys don't know what a BFH is?"

He explains in farmer terms, "A big fucking hole. Rip down what's there, dig a big hole and put a new building up."

We all burst out laughing.

I'm looking out the window and trying to refocus. I am feeding my new tiny little human and think I must be getting dizzy because all I can make out is a building shifting from side to side. It feels like an earthquake and the ground beneath me no longer feels steady. I put my daughter in her crib and draw nearer to the window, still blinking, and there I see Andrew with his truck Betsy hard at work again. He has the chain wrapped around one of the supporting beams in the barn and he's punching the gas forward. Each time the old building creaks forward, then back, then forward, then back, until after the third pull, the entire thing collapses in one big dusty bang and succumbs to the might of the old Ford truck.

The only things we salvage are the foundations and any of the lengthy barn boards that have survived and been weathered over time to a lovely silvered grey. We get to work on constructing a new structure that will house the new office. Neither of us has ever actually worked in an office environment so we start with the obvious. Big bright windows, two large windows on the west of the building where we imagine one day we can watch our children play, two on the north overlooking the vineyard, complete with double doors and a door to the west offering a view of our gravel driveway.

Desks and leased computers eventually fill the space. Andrew and I each have our own workspace with a third for whoever runs the office in our absence. A mini fridge and filing cabinet take the middle position on the far wall and it soon feels like an actual working office; an office away from the house which offers us the privacy that I crave.

The office continues to grow and become fuller, as does our knowledge of owning and running a small business. Our second beautiful daughter, Amelia, arrives just twenty-two months later. She is a lovely and simple tempered baby. She comes ten days late, so I have the luxury of choosing when to be induced. Our only request was "not on a Friday!" I have her on a Sunday and bring her home on a sunny, snowy afternoon on Tuesday in December. I tell her now that she was and always will be the best Christmas present I've ever received. Our hearts and our house are full and this beautiful life we have created continues to grow and flourish.

The convenience of having the office behind the house less than twenty feet out the back door is a Godsend. I have the option of working from the house or office and would be lying if I said I didn't sometimes commute to work in a robe and bunny slippers. Tour guides used to come and go, knocking on the doors while our first baby tried to take her naps. The phone would ring constantly. Now life feels like it has balance. I can skip from the house to the office while we have a part time sitter and back again to the get work done once the girls are settled. I really feel like I have the best of both worlds. I get to pursue my dream of entrepreneurship while not missing a single smile, temper tantrum or first step. I am able to give myself the greatest gifts of motherhood and self-accomplishment at the same time... even though I sometimes relish it through exhausted eyes.

9. The Big Reach

The vineyard takes years not months to prepare for those precious vines. This is another naïve oversight on our part. I had thought it was a matter of simply tilling, prepping the soil and into the ground the vines would go. Dan the Viking farmer (who is actually Ukrainian) plays a crucial part in helping us actually strengthen the soil before we can even think about planting anything.

"You should turn this site into a beach," he laughs, removing his cap and waiving it dismissively at the dry, cracked, stubborn red earth.

The task of soil prep so that we can plant it to vinifera on the back ten is a monumental one. Our plan is to plant Riesling, Cabernet Franc, Pinot Noir and Chardonnay. Weather-hardy rootstocks will do well in our heavy red clay soil. Andrew has taken to visiting the local Tim Horton's early in the morning so that he can listen to the farmers and gain knowledge of what is to come. His fresh face stands out among the locals.

"You must be the new guy that bought the old Palermo farm on King Street," they chuckle amongst themselves.

"You're asking the wrong question," a tea toddling farmer says to Andrew early one morning when Andrew brings up the subject of manure.

"You shouldn't be asking if we use manure, you should be asking what kind we use."

Andrew learns that all of the famers in the area have a secret "manure connection." They only share their manure connection with you if they are sure there's enough to go around.

"Ya, you need a shitmonger," Dan jokes to Andrew and yet again I'm always amazed at how little we know about farming.

Shit is not something I expected my husband to come home excited about. Apparently farmers have their manure preferences. Some favour cow manure, some prefer horse manure and there are even farmers who swear by one-year-old sheep manure if they can find it. Andrew learns that the rule of thumb for heavy clay soil like ours is the more organic the material, the better. We favour "spent" organic mushroom compost and apply ten tonnes per acre every spring. This product is very high in organic material, which helps break down the clay, but is low in nitrogen. We learn that fresh, hot, young manure, high in nitrogen, can push the growth of the vines too hard in the spring, which can create thick woody shoots with inferior fruit. This is all just way over my head and outside my comfort zone. Andrew's in charge of this shit from now on.

The soil in our back ten is heavy red clay. It reminds me of the rusty coloured earth I first saw years ago when Andrew and I took the then-named Trans Karoo from Johannesburg to Cape Town on a three-month journey through South Africa. Small huts and villages sat on the orange dust and the vibrantly coloured powder covered the ankles and feet of the locals as our train clattered by.

It is the colour that surprises me the most when the large D8 bulldozer roars onto the vineyard site and starts levelling out a low point on the back two thirds of the planting site. The topsoil is being scraped aside and the subsoil moved into the vacant strips. Then the topsoil is placed back onto the fresh brick coloured soil. It glows against the vibrant green trees at the back of the long strip of land and I am simply in awe of the contrast; the lushest of green, the palest blue sky, contrasting against the rusty earth, nature's gifts. The memories flood in - another gift - of the journey that took us an incredible thirty-one hours across the desert of South Africa. The herd of elephants running next to the train, kicking up that same shade of red dust with their massive grey heels, the kids who flocked to the station and grabbed the tins of pencils and crayons that we passed through the windows, scuffling, falling and fighting amongst themselves, kicking up the auburn powder from the ground to our train window, forcing us to cover our mouths with our sleeves.

We will have spent a fortune on "under draining" the rows by the time all is said and done. Vines don't like wet feet, so we've gone through the extra expense of under draining every row. The installation of drains under every row from the top of the property to the bottom, three and a half feet below grade effectively lowers our water table. This is so important because we not only want vines' roots to reach deep into the soil to create mineral notes and a sense of terroir, but forcing them deeper allows us to survive a drought without irrigation. Had we bought an established vineyard, the under draining would likely have been done less expensively every third row.

With every invoice that crosses my desk, my anxiety grows. Andrew assures me over and over that "This will be worth it" and "we have to get it right in the beginning" but costs add up and up. Our friends in the business explained to us early on that planting a vineyard wasn't for the faint of heart and now I'm starting to believe them. If I had a dollar for every person that stopped by and told us the joke "If you want to be a millionaire by owning a vineyard, you start with ten million and wind up with one..."

The equipment on the vineyard has also kicked up a minefield of small boulders, some of them the size of basketballs that we spend the rest of the summer collecting and hauling out of the rows. Another

10. Rootstocks

The vines arrive in white boxes. We peek inside and see them - 12-inch-long rootstocks with scraggly hairs dangling off of the bottom. The long, faded red stalks remind me of the warm rhubarb I used to eat from my grandmother's garden.

Andrew is giddy with anticipation. He is talking to the vines like they are babies and I can't control my laughter.

"These little fellas are going right into that red clay and they are going to grow amazing grapes!"

I just pray that Mother Nature is listening and she cooperates! Andrew places the rootstocks gingerly into one-ton apple crates where they can soak up water and swell with moisture for over a week. The vines are desperate for water when they are young and we want the buds swollen before they are placed into the red warm earth that will become their new home for the next 40 or so years.

The top of the narrow, baby rootstock is where the tiny clonal cutting of Cabernet, Riesling and Pinot Noir is. Where the clone cutting meets the rootstock, there is a shiny ball of red wax that protects the joined plant where its future will grow.

Andrew is up with the sun before the hot June day gets away from us.

"Today's the day! It's finally happening!"

He is elated. The vines are ready to be planted. A small crew of five will plant ten thousand vines in the next two days. I first see the laser level when I arrive with two steaming cups of coffee. It is set up at the front and back of the readied field. There is a mammoth John Deere tractor with an odd-looking shiny metal planting device on the back. It looks like a pinwheel with two generous seats on either side and a sharp and straight looking plow between them. Each vineyard helper sits on one side of the pinwheel and they plunk vines straight into the ground as it turns. Man and nature working together to create something magical.

The sharp plow creates a little divot for the vine in the ground and as the pinwheel turns, it sets each vine in place, burrowing it into its new home with precision, exactly four feet apart. The level is connected to an "eye" on the planting unit, so if the tractor jostles left or right, the entire unit adjusts itself, setting the row as straight as a Roman aqueduct. The crew works with phenomenal speed and I marvel at how this bare red field has become something magnificent in just five years and two days... but who's counting...

"Finally!" Andrew grins as he floats around the vineyard and admires the new plantings and his new future. "I can't believe this day has come!"

Quite frankly neither can I. It's been five years! We bought the property in 2002 and the vines weren't in the ground until 2007.

I see the promise of what is to come down the long road. I also see the second mortgage that we've had to take out on the property to plant the vineyard and I secretly wish to myself that they were money trees and not tender vines that can be susceptible to Mother Nature's mood swings. Hail is a very bad word around here and extremely cold temperatures can wipe out an entire vineyard and one's livelihood in one swoop of Mother Nature's frosty spite.

When the vine breaks through the waxy membrane, it wants to grow like a bush. But all of the lateral growth and any fruit clusters are removed for the next two years, forcing the plant to grow upward so that it can become long and thin so that it reaches the top wire. This is called a "whip." It its infancy, we must force the vine to concentrate on forming solid root structures. If the whip reaches the top wire by year two, we can expect a small crop by year three and a full crop by year four.

So, at best, it will be four years before we get a full crop out of this field, and in the meantime after the vines have settled into their new homes, we need to establish the posts and trellising (both of which we have learned have nearly doubled in price over the last few years). Pointy Vietnamese straw hats dot the field. I count thirteen of them working in the hot sun placing nine-foot steel posts in the ground. Four feet is left between the plants. Sixty-two end posts are erected. Then come the seven catch wires that have to be run along the fifteen hundred and twelve-hundred-foot rows.

Then there are weeds, the copious weeds. It is imperative that we keep the vigorous weeds away from the young plants because they compete for nutrients. If the lethal weeds win, our future wines get strangled out of existence.

Some days I see droves of people in the vineyard and my question to Andrew is always the same: "How much is this costing?" Despite all my efforts not to let them, my eyes always seem to roll to the back, although I will them not to. We have teams of hand weeders armed with six-foot-long hoes who spend eight to ten hours a day in the fields removing dandelions, vine weed and Canadian Thistle. They keep these unwanted invaders away from the young vines. I know, I know, all this preparation and maintenance is entirely necessary, but as the days, months and, eventually, years roll on, we bleed money, both earned and borrowed.

11. Smitten

I decide to call our wine brand Back 10 Cellars. It comes to me one night while I am tossing and turning, trying to figure out what to call our wine considering that we need to pay homage somehow to the ten years that it will have taken us from purchasing the vineyard to producing an actual bottle of wine. Putting our name on a bottle just doesn't feel like what we are going for. We would rather the drinker connect with our experience, what is has taken us to get here.

The first of the wines we produce is a mouth-watering Riesling. It is so important for us to make extremely high quality wines from this tiny ten-acre site, so we go the extra steps and hand harvest the tender fruit and make the Riesling with free-run juice and the first pressings. Andrew likes to joke that we make wine the way people used to do hundreds of years ago.

He explains passionately to people, "We don't cut

corners. We aren't afraid to experiment with wild yeast and we actually handpick some of the delicate fruit."

Not inexpensive practices by any means, but it all makes for a premium product. We treat the vines with tender loving care and take extra steps that are more costly, so we ensure it reflects in the finished product. The benefit of starting something from scratch is that you have full and total control of how it's done and with this comes unparalleled quality, however stressful and expensive it is getting to this stage.

The best advice anyone ever gave us about starting a wine label is that people are going to decide who you are from the very beginning and once they decide who you are, the horse is out of the gate and you can't change it. We know our brand has to be good and our wine even better.

We have decided to sell the grapes for the first few years before we get the label up and running and apply for the proper licencing requirements that will allow us to sell wine directly from our retail store.

Dave Johnson, who owns Featherstone Winery, has been with us every step of the way, advising Andrew from the beginning. Engaging a winemaker is like entering into a marriage. We wanted to establish a long-term relationship because we wanted a consistent style of wine year in and year out. Since Dave has mentored us throughout the entire process of site preparation and clonal selection and since he is going to be purchasing our grapes, he has had a vested interest in the quality of our fruit from the very beginning. During that time he also hit his stride as a winemaker and became a dear friend. Every wine in the Featherstone portfolio is outstanding, so when it came to the conversation about keeping some of our fruit to bottle our own brand, hiring Dave as our winemaker was a natural fit. He still got to purchase our premium grapes and we got to test the waters with our brand with him at the winemaking helm. We were afraid to make too much wine at first, because we were worried about not being able to sell it.

To say that I eagerly anticipate the first cheque from our first crop is a wild understatement. I feel like we have been bleeding money for the past five years and I am running out of time, patience and optimism. We've finally pulled off our first harvest and I am dying to know what the crops have yielded.

"What is the tonnage?"

"How much will the fruit cheque be?"

I harass Andrew at every turn.

"Just wait," he begs.

We have an event to go to tonight with Dave, so we will subtly broach the subject and see how we did. I'm doing the math in my head and I'm expecting a cheque at least close to $20,000, which seems more than reasonable for our first harvest.

"Around $8,000," Dave says with a smile, and I immediately deflate in the back seat of the car.

Our first harvest has only yielded between five and a half and six tons. Andrew pulls down the visor above the front seat and I can see him looking at me from the dim light that comes on. His stare looks apologetic, but I am totally miffed.

"The next harvest better be better," I think to myself, and thankfully it is... but not by much.

Only a few short years later, it's time to take the leap of faith and bottle our first Riesling. We name it The Big Reach. It feels apt.

Our girls are now in school full-time, allowing us the creative freedom and time to focus. I feel that we are ready and the grapes are ready and established. It's time to execute that part of the plan and sell our own brand. It's been a decade getting to this point!

The process of designing the label and brand is exhilarating and takes nearly eight months from start to finish. It's the culmination of the past ten years, our dreams and aspirations wrapped up in one big hopeful bow.

The Big Reach is excellent. I know it immediately. I somehow instinctively know that this brand and this wine are going to be all we had anticipated. The label is crisp, clean and unfussy. It isn't pretentious and it reflects Andrew and me - simple, forthright and honest, and a little bit modern. It is very "city folk living on a vineyard," which is exactly what I was going for.

I am gloriously proud of this wine and what it has taken for us to get here. Andrew and I have created all of this out of nothing and this is the first feather in our cap.

Imagine then my dismay when taking this precious bottle of wine and showing it off at a dinner party to hear friends of friends say, "You want to charge how much for this?", "The label is too plain" and, after all our work, "You need to change the branding." Quite honestly, a punch in the stomach would have hurt less. We are crestfallen.

"Chin up," I say to Andrew on the way home from the party.

Andrew looks completely sunk.

"Everyone's entitled to their own opinion," I say.

He rallies and says, "You know you've always gone with your gut instinct, so let's stick with your gut."

I know he's right. I know this brand and this label represent us, and our journey. We are not changing it for anything. It's so easy for people to throw out opinions about something when they have created nothing themselves. They may not even realize that their opinions are hurtful. Not everyone has the proper filter and if there's one thing I've learned about myself and about the plight of entrepreneurship, it is to listen to yourself, to believe in yourself and to do what feels right. Don't second-guess yourself when people without experience throw opinions at you and try not to take their "criticisms" as judgement.

There is a time to listen to people's opinions and a time to value them. I dig deep and I know this isn't the time for the latter, bruised ego and all. There is also a time to know when someone is offering you valid advice and when someone is just being critical. This wine is everything we worked for and I know it's time to trust in myself and believe in our vision. I keep reminding myself that there were also all of the "Ooh" and "Mmm, this is delicious" comments that we have heard time and time again when pouring The Big Reach for close friends and people in the wine industry.

While we're still feeling sensitive about being on the receiving end of those negative comments, as fate would have it, a well-respected wine writer tries The Big Reach at a restaurant in Niagara-on-the-Lake that has just picked it up for its wine list. On his wine anorak blog Jamie Goode gives it a whopping 93 out of 100 points and writes, "This is special. Very fine and beautifully balanced." He goes on to describe the richness of the wine. It's a telling, honest and objective view from a wine expert who knows nothing about us "city kids." He simply reviews the finished product that wound up on his table, in his glass, only about a week after our first restaurant order. The review leaves me feeling validated and beyond elated.

"Are you awake?"

"Hello?"

"Hello?"

"Pick up your phone!"

I'm still so groggy that I have trouble willing my eyes to wake, rubbing sleep from them as I try to focus. I went to bed early last night and now, the next morning, I'm seeing all these texts from Andrew that I've obviously missed thanks to my do not disturb setting. The last one says "Great news!"

I jump out of bed and run into the kitchen, suddenly completely awake and knowing what the texts are about.

We entered the Riesling into a competition, something we vowed to not do often because it usually requires three to six bottles of precious wine and sometimes a hefty entrance fee, neither of which we can afford on a regular basis. This one is for the All Canadian Wine Championships. I reach the kitchen before the girls are awake and Andrew is beaming.

"Oh my God, did it place?" I ask.

"You aren't going to believe it!" he smirks.

I'm hopeful that is has at least placed or received a mention. He pulls up the email and I read it with my hand covering my mouth.

The Big Reach Riesling has won a Gold award for the best off dry Riesling in Canada.

"I knew it," I yell, "I knew it!"

I skip in circles across the kitchen floor. Andrew grabs me by the waist and spins me with his other hand.

"We make an amazing team, Mrs. Brooks!"

He squeezes me close and we dance in the kitchen until our girls wake up. Our first wine out of the gate is a wild success. This wine is special and we knew it from the beginning and now everyone else will too.

12. The Tasting Bungalow

We spend the most amazing Christmas and New Year's Eve in Rome with my sister and her family. We eat copious amounts of pasta, followed by too many gelato stops to mention, and savour crisp walks on cobblestone streets every night after dinner. I love that our off-season allows us to go on adventures in December and January and also that it happens to coincide so nicely with our daughters' Christmas break from school.

The girls prance down the streets of Rome, peeking in shop windows at house-made cannoli and chocolates. I am in my glory looking at the brightly coloured lights that sweep the sky above the Via Del Corso in ribbons of blue, green and red.

We visit Paris, Florence and Venice and the beauty of everything that surrounds us beyond inspires me. Andrew and I duck into every wine shop we can find and eagerly snap pictures of wine labels that inspire us. On our last night in Rome we luxuriate under a heat lamp on a chatter-filled patio at a quaint bistro with red and white checked tablecloths near the Pantheon. We sip on a smooth Chianti and talk about the work that is to come when we get home. We decide to run the guesthouse for one more season and then we can launch our plan for a tasting room at the vineyard.

Upon returning to Niagara from Europe, I am rewarded with piles of mail, a small sacrifice for a three-week vacation, I suppose. In the elastic bundle, I find something rather curious addressed to Andrew.

I put it unopened on Andrew's keyboard, sit at my desk and watch him open it while he's on the phone.

"That's right, yes, we can customize any tour for six or more. Is there a tour online that caught your attention?"

He continues talking to the prospective customer on the phone, trying to sound professional, but clearly distracted by the letter. Andrew flushes and points to it while mouthing, "What the... Oh my God!"

I leap up to grab it. I read it stunned and want him to immediately hang up the phone.

"That's right," he resumes the conversation, "Just let us know when you are ready to book. Bye for now!"

Finally he hangs up the phone.

I'm holding the paper now and can't believe it; it is our winery manufacturer's licence. This piece of paper is precious. It allows us to be designated as a winery, not just a vineyard, and it allows us to sell wine directly to the public from our property. Andrew comes around and takes the paper.

"I didn't expect this so soon!"

"Chris, do you know what this means?"

I do. It means we have more work to do, and another fast approaching deadline.

We filled and filed copious amounts of paperwork the previous spring. We had to dissolve a property line between our house and the guesthouse to amalgamate the entire property and cover in one driveway to satisfy the town and licencing requirements. Subsequent visits took place from inspectors and the like and tweaks and more paperwork followed. We knew it could take up to two years to get our licence so we just figured we'd have another season running the guesthouse.

"Well, should we get the retail space up and running for this season?" Andrew asks, sitting back in his chair looking amazed and unsure.

"Uh... I'm not sure I'm up for this. It's too soon and we have bookings for the B&B already."

I know I sound dubious, but I am so not ready to take on another renovation and we've just spent a fortune on our vacation in Italy. We also only have three skews to open with where we had intended to open with at least five.

"Who opens a winery with only three wines?" I say to Andrew.

He walks over to me, puts his hands on my shoulders

and says, "We do. Let's get to work."

The Riesling, Rosé and Cab Franc are going to have to be the stars of the show.

It's in times like this that I can appreciate that life is all about timing. Sometimes one just has to go with the flow, literally. It's the end of January and if we are going to take advantage of having this unexpected licence this soon, we better get moving. We have four groups booked into the guesthouse for upcoming overnight getaways, so I call them and tell them we are refunding their deposits. I had planned on having at least a year to save for this renovation and now the credit cards will get another workout. We need to be open and up and running in less than three months.

Thankfully I've pinned several tasting room ideas that I really like and I have a vision of what I want the space to look like. I have visions of blowing a hole out of the side of the bungalow and putting in black metal French doors and a large patio, but this will have to wait. We need to get this place open soon and on a very tight budget.

We start by taking down a wall between a main floor bedroom and the dining room. It opens the space up instantly. There is a full kitchen in the back that will stay for now to host future events. We have one curved arch put in to match the others in the entryway and the living room and are more than pleased with the result. This isn't a multimillion-dollar wow winery, this is a grassroots, bare bones winery space and it will have to do. We give the small and humble space an overhaul for the second time since we've owned it. Fingers crossed that people like the wine enough to get past this modest interior.

We know already from our years of running the tour company that the most common complaint of people visiting wineries is being ten people-deep at a tasting bar, standing, trying to get service with fifty other people. We know from the beginning that a seated experience is what we are after. Our space is an unapologetic little bungalow in the middle of wine country. It is what it is and at this point, there isn't much we can do to change it. We call it our Tasting Bungalow and it will be homey, rustic and, hopefully, something that people can connect with. I line the far wall with a long, leather caramel-coloured, tufted banquette and choose a modern cream and light grey paper behind it. Andrew sees the receipt for the seating and looks at me sideways.

"The banquette is the only splurge, I swear," I say convincingly, tucking away the receipt for the specially ordered wallpaper.

I love the way the room looks. I've ordered simple black-sprayed industrial looking chairs with stools to match and have both high and low wooden tables made by Len, one of our fabulous tour guides. He also puts together custom wooden shelves that reach from floor to ceiling. We paint, sand and refinish floors and it's finally done.

The final inspector arrives from the town early one morning and we fling open the freshly painted front door with anticipation. He fills out a paper and hands it to us.

"That's it."

"We're done?" I ask in disbelief.

"Yep, you can open anytime," he says and wishes us luck.

"Oh my God, we need a sign," Andrew suddenly realizes.

It's another task I don't have the energy for.

I hunt down a sign, t-shirts and a computer system that accepts credit cards and tracks inventory. Step by step, all of it comes together.

Putting our wines on the shelves seems surreal. It's taken such a long time to get here. We've been dreaming about this moment for twelve years.

Here are the wines now with their names looking out at us proudly. "The Big Leap," "The Big Reach" and "Rose Coloured Glasses." The lovely faces of the fresh labels looking out at us. It's an achingly proud moment. We tweet pictures of ourselves in Back 10 Cellars t-shirts with the hashtags "Small Vineyard, Big Flavours" and our followers who have come along on our journey, immediately send us warm congratulations.

We open the Tasting Bungalow on a warm Saturday that spring. We are busy. The room hums with happy guests as Andrew pours samples of our award winning wines. I greet guests, pour water and help people with their selections.

"We did it!" the tagline reads on our Instagram picture and I thank each and every person who comes in to support us on our first opening day. A sweet couple come by just before closing and they ask us our story, wide-eyed and eager to learn.

"You're living our dream lives," she says warmly, "You guys are so lucky."

Andrew turns to me and grins. I know exactly what he's thinking. I pour the remainder of a bottle of our rosé into two glasses and we bask in the success of our first opening day.

13. Rose Coloured Glasses

It's a warmer than usual spring evening. The girls are bouncing on the trampoline, the last bit of sun turning pink, catching their hair. Their giggles carry on the soft wind. The end of a busy workday will often find us here, out back behind our office on comfy chairs overlooking the vineyard. Andrew pulls the telescope from the office and we look at the starry night once the sun hides behind the vines.

"Come and look at this, girls," Andrew says as he sets up his telescope and points it at the dimming sky.

I take a peek into it first, and then each of our girls squints up at the sky brimming with stars above. I see Pegasus, Aquarius and Hercules and we point to them while the girls play tag and eat too many marshmallows to count.

Looking back, I don't know how we did it. There isn't a

day when running a farm and a business doesn't require a question to be answered, a problem to be solved or at least a half-day appearance in the office. As a business owner you don't want to spend all your time working in your business, you want to spend the time working on it and growing it, staying ahead of your competition. Andrew and I had spreadsheets, calendars and schedules right down to who worked the morning shift, who had the morning and afternoon shift with the girls, and who was responsible for making lunch and dinner. When the girls were very small, before preschool, the days started at 6:00 a.m. and sometimes ended at midnight with us catching up on our workloads. I remember tapping on my laptop praying for the babes to sleep another ten minutes to get a quote off to a client that I had stalled on for the entire week.

For the first few years, Andrew was the only employee driving the one and only van and hosting the winery tours. I remember watching the newly logo'd van drive down the long drive on its first wine tour. The sun was shining brightly along with our hopes for the future of the tour company. Off he went to pick up our first guests and I beamed with pride. From there we hired more tour guides, purchased more vans and hired more help running the office. Even now, at times, it's hard to know what we need because with each new day comes a mistake we make, something new to learn from, a problem we need to solve or a problem we don't know how to overcome just yet. The cart leads the horse and we continue on chasing and running after the dream of making the tour company more successful.

Nearly fourteen years later, we have the help that we need by way of thirteen part time tour guides, a full fleet of our own vehicles, a fabulous little wine retail store, six wines that sell out in between nine and thirteen months, two wonderful daughters, gold and silver awards, 90+ scores and a nomination for Entrepreneurship of the Year in Niagara. Andrew jokes that in about another ten years we will be living the life that people think we live now. "Wow, you're so lucky!" or "You guys are living my dream!" people say to us again and again. It makes us feel proud, of course, but we just keep putting one foot in front of the other. We certainly see how far we've come but we will always keep moving ahead.

I am not sure that I believe in serendipity, fate or

a predetermined sequence of events but what I do know is that my whole life up till now I have been achingly optimistic about how things will turn out. Of course, there have certainly been times when things were precarious and I worried. Neither Andrew nor I completed our degrees. At the slightest opportunity I would ditch my classes to get on a plane and go somewhere else, anywhere else. I wanted to bite life like a ripe peach and let the satisfying sweetness drip from my fingers. I wanted to savour the smells and sounds of faraway places. I found more pleasure in watching the people around me than in listening to my teachers teach. I didn't want to learn what everyone else was learning and I certainly didn't aspire to be just another resumé floating in a sea of resumés waiting for someone to throw me a life raft.

My life is a sum of all of the wonderful and imperfect parts that led me here, to this place, to this vineyard on this red patch of earth. Each job I loved and loathed, the houses we bought and sold, and the shifts we took at restaurants on our nights off to raise extra money. All of the dreams Andrew and I created together, the ideas, the sleepless nights, the hard work that people see, the hard work that people don't see.

"We did it!" Andrew said ecstatically on opening day.

We did. Our dreams have taken flight. They are no longer ideas scribbled on pieces of paper or contained in conversations in cafés. They are here now in this moment and have grown from our early promises to each other and to ourselves to live the life that we want to live.

Every fork in the road led me to the next one, then to the next, then to this rundown piece of property in Beamsville. Each day of my life drew me closer to exactly where I am supposed to be, where we are supposed to be. My life is a history of my yesterdays and I wouldn't exchange any one of them because it would mean that my days wouldn't end with my girls giggling or my glass of wine, Andrew at my side, on the back deck overlooking the vines and this life that we started on our own terms and with our own determination, from scratch.

The winery and the tasting bungalow continue to be

busy. Most Saturdays Andrew tells guests our story and explains how we've come to be here as they drink in the fruits of our labour. The tour company hums along and continues to grow and flourish, as do our families and our friendships. Our dear friends who helped us when they could have been a huge part of building this life, cleaning rooms at the guesthouse when we needed them cleaned and couldn't afford to hire anyone, helping us paint this mess of a house when we needed it.

We've been blessed to get to know the countless people here in wine country who own wineries and without hesitation simply wrapped their arms around us and mentored us without even having to be asked. Unlike any other business I have worked in, there isn't animosity or a sense of competition among the wineries where we live. None that I've seen anyway. There is a true community here of people who come together to make great Canadian wine and want nothing more than wine from this region to be recognized as some of the best in the world. A common love, common goal and common lifestyle weave us all together.

It's funny to think now that when Andrew and I worked

in Calgary as servers, we would bring Niagara wines to the tables of our guests; Henry of Pelham, Cave Spring and Vineland Estates, among other hugely popular and iconic brands. This past summer Carole, a dear friend of ours hosted a wonderful candlelit dinner on the property of her B&B and we sat with these very wines on our table, with some of the owners of these wineries sitting across from us, many of whom have become our closest friends, Dave and Louise from Featherstone Estate Winery and Daniel and Louise from Henry of Pelham, among others, all raising our glasses and cheersing in the candlelight of a warm, laughter-filled evening.

These vines on our Back 10 vineyard have rooted themselves deeply here. They have become stronger over the years of reaching down into the earth past the water table. It is incredible to me, looking out at the tidy rows of tender fruit that none of this was here when we started our journey. We have created all of this, grown the grapes and our lives into what we wanted them to be. We've watched our children grow alongside these vines, stronger, taller and more beautiful year after year. All of these gifts that we have worked towards fill my life and my heart.

When living in Calgary, I would try to fly home to Niagara as often as I could. A quick respite with friends and family was always something that I looked forward to.

From the Toronto airport, I would hop on a bus that wove its way through Grimsby, Vineland and Beamsville, places that I never frequented even though I grew up just fifteen minutes away.

On a rainy day, on one of those trips, I looked out at the grand properties, green rows of vineyards and the idyllic and charming homes that dot King Street and said to myself, "What a beautiful place. Maybe one day I will live here."

Acknowledgements

This project would not have been possible without the mentoring and guidance of David Johnson and Louise Engel from Featherstone Estate Winery. From the beginning David has consulted on everything including clonal and rootstock selection, vine spacing soil preparation, control of weeds, mould pressure and pests in the vineyard. He is a celebrated and passionate winemaker who has really hit his stride with the Featherstone brand. We feel so grateful to have such a talented and passionate winemaker caring for our vineyard labours. Both Dave and Louise have mentored us, helped us find our "vineyard feet," and answered a million questions. Thank you, thank you and thank you.

A warm and heartfelt thank you to Daniel Lenko for helping us in the early days and teaching us city kids all the "farm speak" we could handle. Without Daniel, we would never had known what a BFH was - and your mom's apple pies will be forever missed.

Thank you to Lloyd Schmidt whose family are pioneers in the wine industry in Niagara and who helped us order the vines and guided us on clone selections when planting the vineyard. Brian Schmidt from Vineland

Estate Winery has also been amazingly gracious in supporting our tiny winery and our tour company. Our dear friend Daniel Speck from Henry of Pelham and his brothers Matt and Paul have always helped with guidance on growing vines and our winery and have been completely generous with advice on growing our brand, thank you!

Richard Harvey from Metro Vino taught Andrew so much about wine when we lived in Calgary. He lit Andrew's passion for wine on fire when he hired Andrew in his bustling wine shop and taught him about wine all the way to achieving his Sommelier accreditation. Thank you, Richard. You are amazing!

A thank you to Rick Small (who has no idea who we are) and to Sal Howell for hiring both of us to work at her fabulous renowned restaurant, River Café in Princes Island Park in Calgary. It was there that Rick Small unknowingly changed the course of our lives and our dreams. Thank you Rick for walking through the doors of River Café and opening our minds to things that we didn't realize were possible.

Thank you to my father-in-law Harvey for being there for us from the beginning and helping Andrew haul, burn and clean up the "back 10" and for being a constant sounding board for us for the past fourteen years - and for being the best grandfather on the planet.

Thank you to all of my family friends who helped us along the way and made the transition to Niagara a little smoother. You came with paintbrushes, cleaning supplies, donuts and smiles and I thank you from the bottom of my heart for all of your love, support and encouragement.

Andrew showers me with love, reassurance and inspiration every day. Andrew, you are a wonderful father, husband and business partner and I am so privileged to be your sidekick on this journey. Ellery and Amelia, my gorgeous daughters, I love you both beyond words. I am so honoured to be your mother. Always believe that anything is possible. Thank you for filling my days with laughter and my heart with love.

BACK 10 CELLARS
EST. 2002

About the Author

Christina Brooks was born in St. Catharines, Ontario. She fled the mundane confines of university in her early twenties and became enamoured of travel. Christina moved to Banff for one summer and stayed for three years. She moved to Calgary for one fall and stayed for ten years.

In Calgary, she became smitten with wine, real estate, entrepreneurship and her now-husband, Andrew, although not necessarily in that order. It was in Calgary that Christina hatched a plan to build a life that matched her dreams. Christina's travels took her to

South Africa, Thailand, Egypt, Mexico, Italy and France and eventually back to Beamsville, Niagara with hubby in tow.

Christina has aspired to be a writer since she was a young child. She has been stringing words and stories together in her mind for a very long time. This is the first time she has ever put anything down on paper and shared it with the world.

Andrew and Christina live on a small, award-winning, Beamsville vineyard with their two beautiful daughters, two cats and too many bunny rabbits to count.

About Back 10 Cellars and Crush on Niagara Wine Tours

It all began in 1999. Andrew and Christina Brooks were working at River Café in Calgary. One wintery afternoon they had a wine tasting with a small vineyard owner, Rick Small from Woodward Canyon, as part of their ongoing wine education. He spoke of his passion for wine, how he had acquired a small parcel of land and started his little winery. The couple were immediately smitten by the idea of doing the same.

What would it take to own their own vineyard? Or was it even a possibility? After all, they weren't rich (far from it), and didn't have backgrounds in farming or agriculture, but they were willing to learn. Although only in their late twenties, they were determined to achieve their dream of owning a small winery.

They worked hard for a year and a half to save a nest egg, living off of only one income and banking everything else they could muster. The hunt began. They were ready and determined to unplug from their lives and take the big leap. They travelled to Italy, France, Kelowna B.C. and South Africa, but these places were too far away and too costly.

It was on a trip to Niagara (where Christina grew up), that they found a ten-acre property in between Vineland and Beamsville. It wasn't pretty. A "diamond in the rough" was an understatement but they could afford it and the house and derelict farm were smack in the middle of wine country.

They drove around Niagara wine country for hours and after getting lost repeatedly decided they should offer a more convenient way of allowing travellers to seamlessly travel from winery to winery. There was definitely a business opportunity here, a wonderful possibility to start a lucrative business. They launched wine touring brand Crush on Niagara Wine Tours and it has been going strong ever since.

The success of the tour company (and a ~~third~~ second mortgage) allowed them to fund the daunting task of renovating a century old farm house and planting a brand new vineyard from scratch.

Christina and Andrew acquired the ten-acre parcel in 2002, planted the vines a few short years later, selling their grapes to another winery for several years, before finally producing their first bottled wine in 2012. They called the winery Back 10 Cellars because it took ten years to get from fallow land to a bottled wine from those once-derelict acres.

Back 10 Cellars' very first Riesling, **The Big Reach,** won a gold award in 2014 for the best off dry Riesling in Canada at the All Canadian Wine Championships.

In 2014 their **2013 Cabernet Franc, The Big Leap,** claimed silver at the National Wine Awards of Canada.

In 2014 their **2013 Pinot Noir, Blood Sweat and Years** took gold in the International Pinot Challenge for one of the best Pinot Noirs in the world.

In 2014 their **2013 Rosé, Rose Coloured Glasses** took a bronze award at the National Wine Awards of Canada.

These were some of the earlier wines entered in competition and Back 10 Cellars has been selling out of each vintage of their acclaimed small lot wines, year after year, since then.

According to the Brooks, getting to this point has been a wonderful and worthwhile fourteen-year journey. There is hard work, laughter, tears and dedication in every glass.

BACK 10 CELLARS

EST. 2002

www.poniesandhorsesbooks.com
First published by Ponies and Horses Books in 2016

CHILDHOOD 6 OF 1

Available as individual, standalone books or as an anthology, **CHILDHOOD 6 of 1**, the first collection of short memoirs from P+H Books, takes a look at childhood from all directions. The funny and moving series of stories features unconventional fathers, unexpected kids, challenging childhoods and plenty of triumph over adversity, expectation, small city limitations and potato sack races.

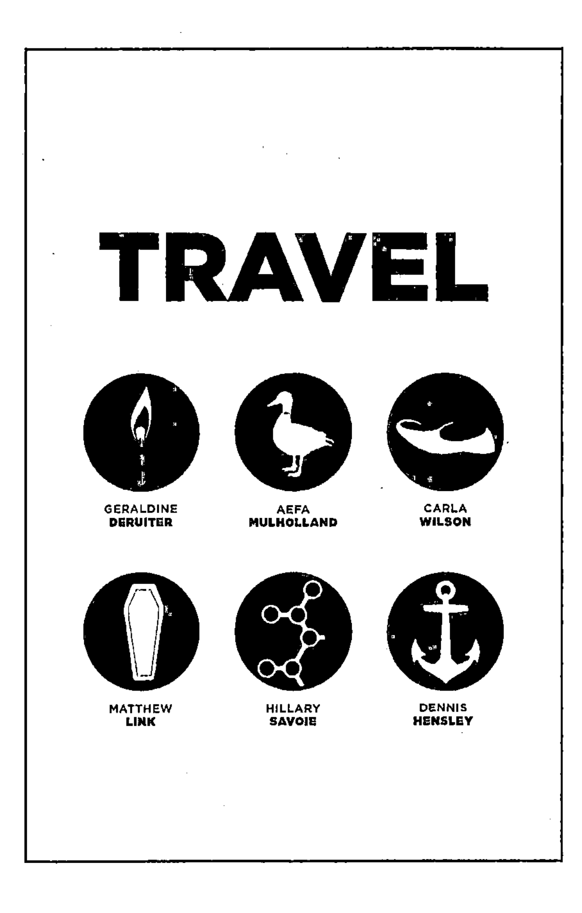

TRAVEL 6 of 1

Available as individual, standalone books or as an anthology, **TRAVEL 6 of 1**, the first series of short travel essays from P+H Books, contains stories that stretch right from the moment of birth to the sometimes surprisingly colourful business of death. There are tales of arrivals and departures, ships that have sailed and ships that have come in. There are stories of festivities perched on the brink of a war zone, on the brim of the Mediterranean and on the banks of the Mississippi.

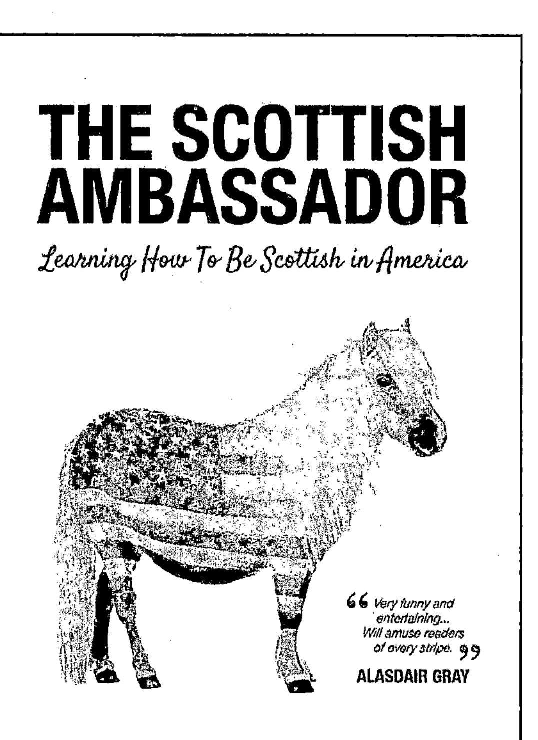

THE SCOTTISH AMBASSADOR

Ex-pat Scot Aefa Mulholland travels around the U.S., persuading bemused Americans to teach her how to be a better Scot. From learning how to do Scottish Country Dancing in Honolulu to attempting golf for the first time on a rattlesnake-infested desert sand golf course in a trailer park in Arizona and learning how to play the bagpipes in New Orleans, she learns about what it means to be Scottish, what it means to be Scottish-American and what it means to be at home such a long way from home.